A Few Crazy Ideas

A Few Crazy Ideas

OR, IN OTHER WORDS, AN EQUATION WITH MANY BEAUTIFUL UNKNOWNS (A COMPUTATIONAL AND MATHEMATICAL VIEW OF THE WORLD)

Petr Rektorys

Illustration © Pavel Kantorek

"Professor, I know that you are highly esteemed, but I really don't envy that mess in your head".

A Few Crazy Ideas
or, in other words, an equation with many beautiful unknowns
(a computational and mathematical view of the world)
Petr Rektorys

Copyright © 2016
Text: © 2016 Petr Rektorys
Illustrations: © Pavel Kantorek
Translation from Czech original: Karel Vavruška, Andrew Chapman
2016
First edition
ISBN-13: 9781537759685
ISBN-10: 153775968X
All rights reserved.

**We are all agreed that your theory is crazy.
The question that divides us is
whether it is crazy enough
to have a chance of being correct.
My feeling is that it is not crazy enough.**

(opinion of a leading physicist Niels Bohr in 1958
to another leading physicist Wolfgang Pauli,
after Pauli's presentation of nonlinear field theory of
elementary particles).

Why should you deal with ideas that are considered to be crazy even by their author? You are certainly a very respectable person, but:

Do you believe that you possess the sixth sense, but you make sure not to mention it to anybody? You are not able to grasp why you had such a crazy dream? Do you suddenly recall something that you couldn't recall the week before, however hard you needed it and however hard you tried? How can your wife listen to what you are saying and know as well the conversation between a friendly couple at our table? Has it ever happened that you remembered someone and in the next minute he phoned as if he knew that you were thinking of him? Have you ever had a feeling of having heard or done something before though it is evidently the first such time? What lies there beyond the edge of the Universe? What is the soul or love? Does exist the aura? Who is God?

If you have ever asked yourself even one of these questions, then you can perhaps find an answer in this book or at least a direction how to proceed in looking for an answer. It could be that some ideas in the book will look crazy, but then you may say that they are crazy enough to be true.

The author's thoughts on the topics mentioned above are of a purely material character, and there are no spiritual considerations using foggy expressions. Quite the contrary. The author looks for the answers from a purely material side and is led by an effort to stay close to the ground. We are usually surprised by the author's simple and straightforward insight.

The author presents the answers to the questions as hypotheses ready for further elaboration and development. In case, you do not like deliberations of the pseudo-scientific type and need a real and novel answer to a given serious topic then this book is for you.

Acknowledgement to the reviewers
(no academic degrees given)

I wish to thank the reviewers for their help in ordering my thoughts so that these may become more readable and may bring as much enjoyment and pleasure as possible. My thoughts will thus contribute to understanding the world and nature as a special case of reality show both around us and especially inside us.

Alice Kostelecká	psychologist and my beloved wife
Martin Kobr	almost my double, but he still believes he is more clever
Viktor Zeisel	a specialist on life, folk philosopher, part-time ICT specialist
František Sedláček	ICT specialist and philosopher
Aleš Přichystal	philosophising winemaker
Marek Šetina	heart surgeon
Pavel Kantorek	illustrator, author of cartoons (Canada)
Karel Vavruška	ICT specialist and philosopher, translator of the book
Hubert Máša	ICT specialist and philosopher
Andrew Chapman	Engineer and Teacher (Great Britain)

The contents

Chapter 1   Introduction – why this book originated · · · · · · · · · · · 1

Chapter 2   Life, soul and other software and hardware · · · · · · · · 7
                 2.1. Computer as a simple model of humans · · · · · · · 7
                 2.2. A simple computer model · · · · · · · · · · · · · · · 11
                 2.3. Forget the electrons and electricity · · · · · · · · · · 14
                 2.4. A brilliant supercomputer in our eye · · · · · · · · 16
                 2.5. I/O device but a fabulous one · · · · · · · · · · · · 18
                 2.6. Operating system · 25
                 2.7. Moreover, what about the human soul? · · · · · · · 26
                 2.8. Definition of life · 31

Chapter 3   Do you belong among the "ten percent"
                 people? Be happy! · 36
                 3.1. How well do we use brain capacity? · · · · · · · · · 36
                 3.2. Subconsciousness never sleeps · · · · · · · · · · · · 42
                 3.3. More heads know more · · · · · · · · · · · · · · · · · 44

Chapter 4   Our good friend Differential Equation or
                 we have our heads full of these · · · · · · · · · · · · · · · · 52
                 4.1. Do you like proverbs? · · · · · · · · · · · · · · · · · · 53
                 4.2. Why the very differential equations? · · · · · · · · · 62
                 4.3. I've got an idea! · 71

	4.4. What dreams did you have tonight? · · · · · · · · · 71
	4.5. Do you like this song and other people do not? · · · 77
Chapter 5	Brain architecture of men and women · · · · · · · · 83
	5.1. What is architecture from the computational view · 83
	5.2. Men and women think differently. Why? · · · · · · 86
	5.3. In comparison with a woman, a man is a simpler system · 91
Chapter 6	You also have the sixth and seventh sense, only you do not know it … · · · · · · · · · · · · · · · · · 93
	6.1. Everybody has five basic senses – really only five? · 93
	6.2. What, in fact, is a sense organ? · · · · · · · · · · · · · 94
	6.3. Sixth sense: intuition · · · · · · · · · · · · · · · · · 97
	6.4. Seventh sense: telepathy · · · · · · · · · · · · · · · 101
	6.5. Moreover, what about the aura? · · · · · · · · · · · · 109
	6.6. Other senses? · 112
	6.7. Molecular and atomic fundamentals of future senses · 113
Chapter 7	Darwin as Euclid · 115
	7.1. What kind of a nonsensical comparison is this? · 115
	7.2. The Law of action and reaction in the evolution of species · 117
	7.3. The Law of action and reaction and feedback system combination · · · · · · · · · · · · · 125
	7.4. What does Jean-Baptiste Lamarck say here? · · 126
	7.5. The influence of will on the evolution of species · 129

Chapter 8   What were you in previous life? · · · · · · · · · · · · · · · · · 133
            8.1. Quackery? No, it is computer
            science + genetics · 134
            8.2. Do older parents have cleverer children? · · · · 136

Chapter 9   I want to be smaller, smaller and smaller,
            to become the biggest in the world · · · · · · · · · · · · · · · 138
            9.1. A muddled heading? No · · · · · · · · · · · · · · · · · · 138
            9.2. Möbius strip · 138
            9.3. Klein bottle · 140
            9.4. Do you like to increase or, better,
            decrease your size? · 142

Chapter 10 Eternal question – what is love? · · · · · · · · · · · · · · · · · 146
            10.1. Nothing supernatural (my apologies) · · · · · 146
            10.2. We are an animal – only a more
            complex one · 146
            10.3. When infatuation moves to love · · · · · · · · · 155

Chapter 11 Faith, God, atheism, and the like · · · · · · · · · · · · · · · · 156
            11.1. Having faith and be a believer,
            there is a difference · 156
            11.2. Everybody has some faith even if
            claiming otherwise · 159

Chapter 12 How the national wealth originates · · · · · · · · · · · · · · · 162
            12.1. How does Ronald Reagan relate here? · · · · 162
            12.2. What is at the beginning? · · · · · · · · · · · · · · · 163
            12.3. Too naive? Too unrealistic? · · · · · · · · · · · · · · 167

Chapter 13 Epilogue · 168
            13.1. What to add? · 168
            13.2. Immodest wish · 168

Chapter 14 Glossary · 169

Chapter 15 Peer reviews · 175

CHAPTER 1

Introduction – why this book originated

Motto:
We are all agreed that your theory is crazy. The question that divides us is whether it is crazy enough to have a chance of being correct. My feeling is that it is not crazy enough. (opinion of a leading physicist Niels Bohr in 1958 to another leading physicist Wolfgang Pauli, after Pauli's presentation of nonlinear field theory of elementary particles).

The book originated as the result of my lifelong search for answers to questions that humanity have researched from time immemorial and to which the philosophers, scientists, poets and others have tried, and are trying, to find answers.

My questions related to life, the world, the universe and everything ☺, were usually rather dry. I have tried however that they are directed towards the core of the matter. I will do my best to ensure that reading this book is as entertaining and pleasant as possible, even though it will touch upon the serious topics in life.

Some of the fundamental questions which I began to ask myself at the age of about twenty, were:

- Can I create a model of this world (in any way, on a computer or otherwise) such that it is a hundred percent accurate?
- What or who is God when the majority of people around me believe in it or Him?
- Can I research God's properties?
- How is it possible that a great part of the cleverest people on the earth (scientists or philosophers) have turned themselves towards God in the final stages of their life, even though they were non-believers or atheists their whole life?
- What is love when it can hold us in its grip so firmly? ☺

In this book, I will endeavor to answer these and some other questions. From a deep scientific research point of view, my answer will be perfunctory (I am not a scientist; however, the effort to use scientific methodology is exhilarating for me), and I do not mind if my answers or opinions are perceived as provoking. What I have mainly in mind is to stimulate thinking about these questions from other directions than until now. I hope it will inspire someone to either passionately cut my ideas to pieces thus giving me valuable opponency, or elaborate the ideas, provided he or she can find a grain of truth there, thus giving the ideas a chance to prove themselves and show their contribution.

I have been trying to think about these questions even though the chance that I can find the answers was not high. However, just the mere effort to ponder these questions was a cause for joy. These questions have excited my mind over a long time, and I could not get rid of them. I never was (nor am I now ☺) a top computer scientist nor am I a top philosopher nor a top physicist nor a top mathematician nor a top theologian to qualify me to find the exact or at least an acceptable answer. Instead, I have been researching these questions as an entertainment and enjoyed the feeling that perhaps something may occur to me which I could call (at least a partial) answer.

A FEW CRAZY IDEAS

This is why this book originated – I have tried to critically evaluate the answers that occurred to me in such a way that I may be able to communicate them in good conscience to my immediate circle, without being accused of alcoholic intoxication. I would like to present here the result of my research of life, the world, the universe and everything ☺ and hope at the same time that someone may be tempted to engage in something from this book. Were this the case, I would be happy and would have the feeling that I have not lived in vain.

The basic rule for finding answers and hence also for this book is:

The only limitation I will respect is that a given idea should not contradict the laws of nature.

You could argue that there are certainly laws of nature we do not know. Yes, this is true. Hence, I would like to expand this basic rule to mean "laws of nature known until now". In no case do I doubt that people will leave no stone unturned and discover new laws of nature but I do not want to get into questions like "can something move faster than light"? Yes, this is indeed a thoroughly legitimate question but it is a consideration beyond the known laws of nature and that this I do not want.

I would like to show my approach in an example which inspired me some long time ago and provided me with a "thinking model" about what is truth and what is not. This idea was a revolutionary one for me at the beginning, and I felt a bit cheated that what I learned in the elementary school in geometry is actually not true.

Who of the normal human beings would ever doubt that such a well-known and easily verifiable fact that the shortest distance between two points is a straight line (this is one of the fundamental laws of Euclid[1]

---

[1] **Eukleidés** also **Euklides** as well **Euklid** (approx. 325 B.C. – approx. 260 B.C.) was an ancient Greek mathematician and geometrician. He lived most of his life in Alexandria in Egypt.

dating back about 2,300 years) could possibly be not (quite) true ? However, it holds true on a very limited scale, a meter or a kilometer, but on the scale of "the earth" (the surface of a sphere) the theory starts to falter — simply put, the shortest distance between two more distant points on the surface of the earth (thousands to tens of thousands of kilometers) is no more a straight line but it is a curve (part of a circle) as we have to move on the surface of the earth. Airlines provide for their passengers onboard world maps with the flight paths. These flight paths are "kinky and not straight" and make the passengers wonder why the airplanes fly in such stupid ways. They could save both the time and fuel by flying straight.

In reality, the airplanes follow the optimal flight paths (the airlines do take care about this) and the paths take into account not only the distance on the surface of a sphere (the earth[2]) but include even the known wind currents (direction and speed) at various altitudes. It is just this view on the "shortest flight" of an airplane on the earth which reminds us of the "shortest distance" in the universe where, because of the theory of relativity (among others, variable gravitation), the shortest distance between two points is not a straight line. It certainly poses a problem for an ordinary mortal to imagine space warped by gravitation; however, imagining an airplane struggling against changing headwinds or sidewinds, but rising two kilometers higher may make the wind more favourable, fuel consumption will drop, the flight will be faster — almost everybody can, in fact, imagine this. ☺ There is nothing to write home about such an imagination but it may be surprising if we experience it for the first time.

That is why I believe that at least one from the list of ideas presented here will be challenging for our new take on the world.

---

[2] the shape of the earth is not a precise sphere, it is rather a "spherical potato" — the expert's name for that form is "geoid" — see https://cs.wikipedia.org/wiki/Geoid

A FEW CRAZY IDEAS

"Well, the good news first. There will be an entry about you in the medical books".

Illustration © Pavel Kantorek

I believe there is enough space for new discoveries, solutions and takes on the world within the known laws of nature. The point is not to be afraid and apply an innovative approach even though we have no idea at this time how to implement something – for example, to manipulate not only individual atoms (though scientists do it in the laboratories), but mainly proceeding just a small step further – to the level of the particles making up atoms – protons, electrons, neutrons, quarks, et cetera. There are many of these, and some have really strange names: one quark even bears the name "Strange" ☺, the scientific name being "s quark" where "s" stands for "strange". In the beginning, not even the scientists could understand the peculiar behavior of this particle and that is why they called it so.

However, why should it be impossible to learn to manipulate individual particles? Nature does it regularly and, after all, we could not even fly about 110 – 130 years ago³. Just about 50 years ago we could not look through a microscope and observe the individual atoms (but yes, we could see the bacteria). Why then shouldn't we in 50 years use telepathy rather than mobile phones for understandable human communication when perhaps the principle lies at the molecular or atomic level (see chapter 6. – "You also have the sixth and seventh sense, only you do not know it …") and it could lie just a bit away from our understanding and technology? We should always look at these issues as if nature knows how to do it, and if humanity can learn from nature and replicate it, humanity can do it as well.

That is why the fundamental rule governing the considerations and hypotheses in this book is:

Everything not contradicting the known laws of nature is possible.

We only need to imagine it and then prove it. Let us go further …

---

3 First flight of unpowered airplaine heavier-than-air ever – Otto Lilienthal in 1891, first flight of engine-powered and manoeuvreable airplaine ever – bros. Orwill and Wilbur Wright in 1903.

CHAPTER 2

Life, soul and other software and hardware

2.1. Computer as a simple model of humans

I had amused a couple of people when I told them truthfully why I devote myself to computers: I started to devote myself to computers in order to better understand humans. ☺ In fact now, same as 40 years ago, a computer (any modern computer) serves me mainly to understand how a human is perfect and advanced in comparison to a computer (any computer existing today). ☺ My compliments to nature for this creation.

Computers were first my hobby, and then I began to earn my living in computer applications, and in the end they became my destiny (not in the sense that one fell on my head ☺ and caused this book to happen). Computers got me to computer science that is the science of information processing. Computer science is also a way of understanding what happens around us. This is what has become my destiny. I fulfilled my original mission – I started to understand how things around me play out, not only things and processes dealing with humans but using computer science, I started to understand some things and processes in the universe in general. Some may say this is an exaggeration; however computer science seems to be the key to understanding the world.

It would sound appalling but do we know at all how such a most basic process of forming a molecule of H_2 from two atoms of hydrogen ("H") originates and proceeds? This happens when two hydrogen atoms meet and one "says" to the other "hey, I wish to bond with you" and the other one nods for an approval and the bonding happens. These two hydrogen atoms exchange information. How? It will be a while before we can answer this question. Perhaps the answer is already known or suspected by some top physicists and mathematicians solving problems only they understand ☺ - I mean quantum physics, string and superstrings theories, and other – but perhaps not even they know the answer. They just assume it "happens somehow". But let us return to humans.

We can consider a human being as a biochemical system (processes run on a biological basis) or a static-dynamic system (skeleton, muscles and others, and processes, how it all functions, that "it" can move and does not fall or perhaps "it" stays unmoving). We can certainly find different models and views of a human as a system. One such view is certainly the computational system. Perhaps this is the most important system. Why ? Because I am so arrogant and claim so? ☺ No – I feel it primarily because this describes what we usually call our soul or spirit – compare the proverb "A sound mind (software) in a sound body (hardware)".

It is the very soul or the nervous system (brain, spinal chord, nerves) which control not only the biochemical processes but also enable humans to stand, walk and run, and control additional processes. Moreover, evolution ensures the development of other features for survival. A human (or any creature) will grow, for example, a fin (the soul or software recognizes the need of something like this for survival and instructs the body to create "something like this"). In the beginning, the fin is small and unimportant but after a sufficiently long time it will become a normal fin (or other limb or body part needed for survival – all this, of course, happens only if the creature can adapt quickly for survival). Hence, I suppose that the computational model of humans (human soul) is primary, and the other models must be subordinate.

When I watch the scientific-popular releases or read articles about modern scientific brain research on the activities controlled by the brain, I feel rather sad. We use external methods (CT – computer tomography or similar) to find out which part of the brain is active (that is, we can measure that a small piece of brain tissue, say a volume of one milliliter, is active, which could mean we measured the electrical or magnetic activity or heat only). It appears as if we tried to understand the fundamental functions and computational processes of a processor, consisting of a several logical units with roughly a billion transistors, by observing it and stating: if this computer game is played then the regions at bottom right and top left has got warmer, which means this is the region for playing games. This is really a big piece of absurdity.

Were I to try and offer such a procedure to those professionals, who design and manufacture processors, I would look like a fool or clown. Similar nonsense would be if I were trying to measure traffic among the individual interconnections between the processor and the memory (there are 64 such interconnections in a 64-bit processor) and tried to understand memory management in an operating system (Windows, Linux, and others) which runs on the processor.

In fact, we do not employ a different approach to studying the brain at present. Completely missing is the research "from below", that means understanding the basic functions of neurons (this is a task for one whole generation) and not only its biological way of functioning (that is how mathematical and computational processes are done at the biological level) but also which mathematical and computational functions they in fact are. I can try to explain it by way of an example – it is a well-known comparison of a computer to an idiot who does not know anything else except adding two numbers fiendishly fast (a billion times a second, for example). This is, in essence, true. However, for a long time already, computers can evaluate the sine function and for this they need a mathematical procedure that is put into the processor first as a design and later made into silicon. A processor thus equipped may be used to compute a 3D model of a scene in a computer game (the model is meanwhile hidden in the computer memory thus is not seen anywhere) and only later do we undertake how to present the scene to a human. And I keep a detail for the end: a design of the story of the game is orders of magnitude higher in complexity than merely computing the sine function.

I have described, in rather simple terms, the path from the brain neuron to its activity for the existence of a creature (human or animal) in nature. Why do I mention all of a sudden an animal and not just a human? It is because of something possessed only by a human – abstraction. Thanks to it, a human designs (invents) what he or she may want, and then finds a way how to achieve it and then implements it. And along the road it takes care about nutrition (stocking up on energy and material) as well as survival (own security). Such an abstraction (see Glossary) is something that today's machines and animals cannot do, and even a human has difficulties to describe what it is. Well now, all this must be taken into account in trying to understand how the human brain works. The CT method could be only good to let us know which part to cut out, try to heal, or to see what is wrong there (details are usually not known, only that the situation differs from the norm).

It is very difficult to describe the soul or the spirit of a human and that is why we, at present, limit its description to a biochemical-physical aspect only (for example, which parts of the brain are filled with blood when doing some definite activity and hence which part of the brain does what; however, quite without an explanation what happens inside those parts – which computational operations) or psychological aspect (different levels of the soul and logical aggregates, their malfunctions, and similar) or abstract philosophical aspects (will, love, hate, et cetera) and links among them. All these aspects are right, but none goes to the heart of the matter in the sense that none examines how the soul functions.

As if I can hear it – the cheek to call such a sacred entity as the soul so mundanely and mechanically, after all we stand no chance to understand and research this complexity, and so on. Yes, this is the technological view of something so precious as the human soul (even the animals have souls but not so complex and convoluted, and, mainly, they have no consciousness[4]). May I invite you now to a short excursion, before starting to describe how I understand the computational model of a human being (human soul), an excursion describing the basic functions of a computer. Please be not scared and continue reading. I will try to use a "human language" and no IT lingo. ☺

2.2. A simple computer model

A computer is a machine for information processing. There is nothing world-shaking in this finding but it may be good, however, as a reminder. The computer simply gets some information, it may be a message, in some way ("input" in IT parlance), processes it somehow and passes them on ("output" in IT parlance). Let us imagine for illustration that such a message may be similar to a donor text message (DMS – Donor Message Service) we are sending from our cell phone. Such a DMS can, for example, look as follows: DMS ALZHEIMERSTOP[5]. I am giving a command here to donate one dollar to the charity project ALZHEIMERSTOP. Information can be rather complex, in the sense of being much more extensive and more richly structured than the two words mentioned above. Nevertheless, the fundamental information to be processed in modern digital computers is "yes" and "no". The computers denote this as one and zero. These are two basic states, and thus we have what is called the binary notation. The reason is that it is rather simple to physically distinguish between two states – for example, the electrical current either flows or does

---

4 see Glossary
5 This is real–life example – http://www.darcovskasms.cz/projekt-678/alzheimer-stop.html

not flow, a face is either smooth and reflects a laser beam or has a hole and does not reflect this beam, and so on).

A clever gentleman of the name John von Neumann[6] designed the basic functional model of a computer (the architecture) in such a way that we have a "storage" for information (memory) and additionally a place where the information is processed (processor). It works in such a way that the processor takes the information, for example two numbers, from the storage and processes them as per instruction – for example, it should add them – and so it adds them and puts the result back into the storage. This instruction is the program and the information being processed is called data. Data being put into processing are the input, and the result is the output. Such computers are called digital because they process digits. **All numbers, characters, pictures, sounds, et cetera, are ultimately stored as binary digits in digital computers**.

Practically almost all computers we know today work in this way. Therefore it is not even mentioned that they are of the von Neuman architecture. It is still work-in-progress in the laboratories on photonic computers where light is used instead of electricity. Photonic computers have an advantage that they can simultaneously use light of multiple wavelengths in one fiber optic conductor (for example, 8 or 64 wavelenghts so that 8 or 64 different binary numbers can be processed at the same time) unlike an electricity conductor which either conducts electricity or not so that we can use it to store only one piece of data (0 or 1). There is also work on the quantum computers that are, however, in very early stages of development and production.

Comparatively little is remebered today about the so-called analog computers, and it is even less known that these were the only computers before the digital computers (which could process exclusively zeros and

---

6 **John von Neumann** (Dec 28th, 1903 Budapest, Austro-Hungary – Feb 8th, 1957, U.S.A.) was a mathematician, who heavily influenced quantum physics, functional analysis, theory of sets, economics, computing sciences, numerical analysis, hydrodynamics, statistics and many other mathematical disciplines. Von Neumann architecture of digital computers was created in 1945. This concept is used in nearly 100% of today's computers.

ones). What else could then the analog computers process if not zeroes and ones? They could process continuous variables that, however, were not cut into pieces (digitized). Examples of such variables include, for example, electrical current or material strain or temperature of a solid body.

Any value is represented in a digital computer, unlike a analog computer, by zeros and ones – if we need to record 1,000 possible values (for example, the brightness of one pixel in a digital camera), we record these values in steps of one, that is 1, 2, ..., 1000. We cannot, however, record the intermediate values such as 10.5 or 50.333. If we want to describe this feature in more detail, we must refine the steps and describe a value of, say, between zero and a million, thus we can catch the value more precisely though of course we are still limited by steps of one.

Analog computers do not use zeroes and ones but do use a physical phenomenon, for example, electricity. We create a circuit (conductors, resistors, and other), let electricity flow into it until a stable state is created. The solution of the task is the value(s) at selected junction(s) of the circuit (for example, we can measure voltage or current as the result) and we are done. A digital computer must create a model using zeros and ones, must have a precise mathematical procedure for the solution (an algorithm) for any given task and then it computes the solution using that algorithm. Any analog computer works as a material model of the real world where physical variables – electricity, liquid, pressure, et cetera – are used for the solution of the given task. We can be sure that if we break a wooden stick, everything goes according to exact mathematical and physical laws[7], and the stick has no need to learn anything and knows exactly how to break itself according to those laws. ☺

I can almost hear it: why do we make it so complicated when the analog computer is much simpler, more elegant, and the gut feeling tells us it is even faster in some cases? The answer is simple: at a certain level of

---

[7] It is a differential equation, which describes this phenomenon – static stress by simple bending during breaking a wooden stick

complexity it is almost impossible to set up a material model as an analog computer. Thus, for example, modelling a car body on a digital computer is much easier and more precise and testing the strength of it a thousand times, while destroying it, is faster and cheaper than making a thousand material models and subsequently "destroying" them.

2.3. Forget the electrons and electricity

We can, in principle, imagine another form of a computer than digital and analog ones — it could be a quantum computers. I will not go into details here as they are beyond the scope of this book — indeed, at present there exist systems that simulate simple quantum computers only, and there are no real quantum computers. The other form of a computer is the (bio-) chemical computers which can process information recorded in a (bio-) chemical way. The prefix "bio" is in parentheses because this chemical information (data) carrier can be either "only" chemical that is at the level of molecules of chemical substances and their chemical reactions (that is, "non-living" matter) or at the biochemical level, that is using the biochemical structures of matter (that is, "living" matter). Theoretically, a computer based on the amino acids can be the result of an independent discipline in mathematics and computer-science including corresponding mathematical operations. Amino acids could be information carrier but with substantially many more states than just the two, zeroes and ones, as it is in today's electronic computers. If we use an amino acid as the information carrier, we could call the information thus captured an "aminoformation" = AMINO acid + inFORMATION) and the logical components based on amino acids could be denoted as "aminonics" = AMINO acid + techNICS (the equivalent of electronics). Moreover, storage of the infomation in a molecule of amino acid would be very stable because the molecule itself is very stable.

At present, modern electronics struggle most with the heat generation (today's chips would burn within seconds without the a heat sink device) on the one hand; and with so-called crosstalks on the other hand. The crosstalks are states when the electrons break through insulation and cross over among the individual lines on the chip due to the the miniaturisation. Thus we cannot be certain if the original information "1" is still "1" or has been changed by crosstalk into "0". The smallest dimensions in modern electronic circuits have a path thickness of about 10 nanometers (in memory chips – as of January 2015) which means that 8,000 such lines placed one-by-one would be the thickness of a hair! In contrast to this, the amino acid molecule, as the basic logical element of aminonics, would be an extraordinarily stable element (a molecule is stable by its nature) though it is much smaller than the smallest transistor in today's electronic circuitry. It would be necessary to design a way of communication among the individual amino acid molecules for the complete function of aminonics logical circuits. Obviously, it would be convenient to use a suitable molecule for this communication. Such a molecule would react with the amino acid molecule and thus would read or write the aminoformation. I can imagine this to be an analogy of the peptide bond.

The operations on aminoformations would be defined mathematically in the same way as today: we can use the basic operations to add two ones – in the binary system: 01 + 01 = 10 or in the decimal system 1 + 1 = 2. Modern processors have 100 to 200 basic instructions collectively called the instruction set. One such instruction is the "addition of two numbers". Chemical reactions (reactions of amino acids) using aminoformations could theoretically be distinctly more effective (more complex and much faster) than today's primitive binary operations with the primitive information items like 0 and 1. A chemical reaction goes so fast that it may stump our modern designers of electronic components[8]. The reac-

---

[8] It is virtually impossible to find even an estimate, how fast (timewise) is the chemical reaction not only of two aminoacids (i.e. how long it takes to create peptide bond between two aminoacids), but even between two hydrogen atoms (H) to create one molecule H_2, or how long time it takes to create one water molecule (H_2O) from two hydrogen atoms and one oxygen atom.

tions could, in essence, process instructions that are inborn in them (a chemical reaction of two substances can, in fact, be an instruction for such a biochemical computer). In comparison to these instructions, it would appear that modelling of a process using zeroes and ones is very inefficient.

Why do I mention it? Because a human is such a biochemical computer where the amino acids form one of the basic elements. It is an extraordinarily complex computer, which is resistant to errors and attacks (self-healing capability). My compliments to nature for this piece of art.

2.4. A brilliant supercomputer in our eye

Perhaps the most beautiful and most exciting example of what I am describing here I have born in my mind since roughly 1986. I read an article at that time in the Byte magazine, which, unfortunately, is not published anymore, and I immediately copied and filed the article. Described there are the fundamentals how the biochemical-computational principle of humans could work. There are still chills up my spine even today when I recall this as we have not got too far in this direction. The description concerns the functions of the retina of the human eye.

At that time, a scientist tried to elucidate the complex goings in the human body using the example of human eye retina. I will try to describe briefly but precisely what that article had about the retina of the human eye (the literal quotation would not be so readable for non-IT people):

Each cell in the retina of the human eye performs a complicated task that could be mathematically described with a system of 100 x 100 non-linear differential equations. The most powerful computers of those times (author's note: the year was 1986) **were of a size to fill a gymnastics hall and had the power consumption in tens of kilowatts and solved this task in about three minutes. One cell of the**

> retina solves this task twenty-five times a second. There are about 100 million of these cells mutually interconnected and cooperating to create an image.

Wow! The end of the quotation. Even now, after thirty years, I can feel a chill up my spine, as we have progressed closer, yet we do not know how it all works, not only at the biochemical level but not even at the level of basic processes (mathematical processes and computational algorithms derived from the processes).

This is a quotation which evidently describes this matter with sufficient precision and that is why I do not want to look for possible inaccuracies. We do not know, for example, if the function of a retina cell corrresponds more to an analog computer, while we (the humans) try to model its function in the digital environment. This can be very ineffective though mathematically precise. We can even dispute the fact "precise" — why would it be proper to use just the 100x100 system of equations and not 1000x1000 or 10x10? Even from this point of view, they obviously refer to an approximate description of reality, so common just in non-analog systems (i.e. non-continuous — called in mathematics as "discrete"). A mathematical task can be solved of course by an analog computer. However, we (the humans) are technologically not able to create an analog computer at the level of the retina cell. We should, however, use such technological procedures (mathematical — computational — biological ones) for the understanding how such a cell functions though we cannot build it.

Even if we admit that the processes in the retina cell function on the principles of an analog computer, we encounter some "digitization" or, maybe, "discretization" on the way between the retina to the brain. The digitization concerns objects recognition in complex 3D situations (we can see a part of an object only — for example, women's hairstyle), object's mapping on the known abstract notions (after objects recognition), situation evaluation and formulation of conclusions for further activity. All of this cannot reliably happen at the analog level because we are dealing here with discrete abstract notions and operations on them. An analog approach can hardly model such an abstractness.

We do have much more respect for the eye now, don't we? In order to enhance the understanding of the performance of the human eye, we expand the performance explanation of this biological computer by mentioning a specific task. The task includes the cooperation of the eye with the brain. This task has stumped humanity until today even on the algorithmic level, i.e. the individual steps in solving the task: object recognition (see the previous paragraph) – what the eye caught and made up into a picture (as a photograph) but in this picture we can not only recognize what we see but we can involve other relevant information (emotions, intentions, social role, et cetera) and other relations to this computational model of the real external world. And all this is done in a matter of tenths of a second from the time we take a look at the scene.

2.5. I/O device but a fabulous one

Just a few little notes concerning the parts making up a computer: in addition to the processor (here information is processed) and the main memory[9] (which stores the information prior to and after the processing), there are devices for contacting the external world (so-called peripheral devices, in short the peripherals or I/O devices). These devices may serve for the transformation of a non-digital object into the digital form – for example, a digital camera creates the image of a landscape in zeroes and ones, a scanner takes up (digitizes) a picture drawn by hand on a paper, a microphone picks up a song and makes it into a string of zeroes and ones.

Conversely, some I/O devices serve to transform zeroes and ones into a form understood by human senses – a display (creates an image or text but only in "soft" form as it can quickly disappear being replaced by a new

---

9 Other denomination: internal memory or RAM (Random Access Memory – it is possible to write into any RAM memory cell or read from any RAM memory cell).

text or image), a printer (creates a "hard" copy of an image or text by putting marks on matter – ink on paper), a big hit of the present time (2016) - a 3D printer (creates a three-dimensional shape from the zeroes and ones in the computer), and, in a way, the engine of our car that is controlled by a computer (yes, the control unit is the computer) is an I/O device of the system "control unit and the engine". We should not forget one additional type of an I/O device, namely external memories.

As such, the topic of memory as "information storage" is very important. There are many types of memories – for humans, the scientists are agreed to distinguish among a short-term memory, medium-term memory and long-term memory[10]. The short-term memory serves for immediate response, as for example in life emergency, car driving, et cetera. The medium-term memory is the first stage of information filtering and perhaps already the first stage of coding and generalization from the short-term memory. The long-term memory serves primarily to maintain the experience for future usage and is probably placed exclusively in the subconsciousness. Similarly, in computers we find the operating (internal) memory that is usually the fastest one in a computer in so far as information processing is concerned but usually also one with the least capacity because it is extraordinarily difficult to manufacture. In fact, there is one more type of memory, which is even faster and smaller in capacity, the intermediate memory of a processor, so-called cache. However, I do not want to complicate matters in the book because it is just another type of main memory where the programs run.

We can be even more precise: processor registers are in fact also memories but these are a built-in part of the processor while all other memory types are used by the processor to store intermediate results "outside the processor", and thus we will not consider the registers as memories proper in the computer architecture. Furthermore, a computer has also the external

---

10 Some sources quote other human memory types – sensoric (also: ultra-short), which is the memory of individual senses for information acquisition (optical, acoustic, gustatory, et cetera), which is sent at once to the short-term memory and consequently overwritten by new information based on work cycle of the specific sense (for example, it is approximately 1/25 of the second at the retina cell).

memories of different types based on their capacity to store information and their speed to retrieve information and transfer the information to the main memory – a disk memory is such an example. The processor instructions can only process information stored in its main memory (and from cache and registers), it picks up from there the needed data, moves them into the registers, and processes them and puts them back to the main memory. In comparison to the main memory, an external memory is always slower in information retrieval and usually disposes of larger capacity and also, information is not deleted at the moment we switch off the computer.

We will see below that the way how information is stored in the operating and external memories is fundamental for both people and computers. The ability to store the maximum information and at the same time be able to process it fast enough is important in practice, so that humans may survive, and the computer may solve the task in sufficient time. As an example, it would be foolish to have a weather forecast for two days ahead, if that information took a week to compute. It may be precise (hopefully! ☺) but we will have no use for it as this is in fact already history. In a similar vein, it is essential for human survival to decide whether to fight or flee in a given dangerous situation. If the human does not decide sufficiently fast, he or she will die which is contrary to the basic principles of the system called a human. ☺

A typical computer with its peripheral devices possesses one fundamental feature: information processing (input and output) has no feedback. The computer simply reads the input data from something (disk, keyboard, scanner, camera, reader, measuring device,…), processes them according to a program, "spits" them out somewhere (disk, display, printer, or other) and that is the end of the story. ☺ It does not do anything else with them. Perhaps it may be better for a robot instead of a computer to compare a machine and a human being because what we need to have is feedback from the external world; that is to say we must collect input information (using the senses or sensors) and respond to the environment

A FEW CRAZY IDEAS

(to control the system using the information based on the data obtained from the environment).

Such a system is called a cybernetic system, and such machines (perfect robots) are sometimes also called "Cyborgs" (CYBernetic ORGanism); however, I prefer the word "robot", and it is even used more frequently. By the way, do you know that the word "robot" is a donation to the world from the Czech language (the author's native language) and that there are several other such Czech words in common use – robot, dollar, howitzer, pistol, polka? Aside from the Czech speakers, few other people know that "robota" was the hated forced labour of serfs on the fields of their masters and so that the name "robota" for a machine doing such work is exact.

Unlike a computer, a robot works nonstop (if switched on and with sufficient energy) and continuously creates feedback between input and output and all this is done in harmony with the internal conditions and demands for the robot functions. For example, such a robot may be given a command to move to a certain place. It is extremely difficult to check the robot's position with the optical signals as we humans do it using the eyes (a robot's camera snaps a picture and the robot's control system evaluates whether this is the place or whether the robot approaches it) and thus it is much simpler for the robot to get its bearing using a technologically simpler way – a GPS[11] signal, for example. The robot thus gets its bearing in the space where it is and where the destination is and subsequently issues a command to its engines (often electric motors), that control its movement (it does not matter whether the robot moves on wheels or on foot). If the movement is in the direction to the destination, everything is ok, and the robot continues on its way. If not, and the robot finds out it has strayed off its route, a correction is made, and new signals are issued to the engines.

Humans have in principle a similar feedback system, it is just more complex and more precise, only the movement of a human to the required

---

11 GPS (Global Positioning System) – satellite system, which can set the position of sensor on the earth with the precision of one meter / yard (in the case of public version of GPS – for car navigation, et cetera) or centimeters / inches (military version of the GPS).

destination is based on surrounding information captured by eyesight (completed by acoustic, tactile and smell information, of course, thus creating the model of surrounding), instead of GPS. The task just described is based on neighborhood capturing by optical sensor (the eye or the camera) with subsequent picture evaluation at the level of picture's details. Such a task is beyond the computing power of modern robots even though it is "only" a determination of a position in space. In its way, such a procedure means squandering the computational power of the brain, but it is correct from the system view because humans do not depend on additional system supplying the positional information (GPS). Humans are thus fully autonomous.

Another example is when a human receives information into the central control system (subconsciousness and consciousness) about water shortage in the system of biochemical processes (thirst ☺). Then consciousness issues a command to the eye (I/O device – input) to find a source of water (a glass of drink, creek, et cetera), and when this is found (a rather complicated task), it instructs the hand (I/O device – output) to reach for the glass of water and move it to the mouth (I/O device – input) and pour water into the human storage tank (we drink). It would appear that the task is completed but by far not yet – a mechanical-chemical process is triggered in the tank (stomach) to deliver water to all human cells to sustain the basic life functions.

In principle there is no difference between a (bio)-robot and a human from the viewpoint of information processing (a computer-science viewpoint). The big difference, however, is that humans (and animals) have life that is a specific category (self-awareness as a definition of self against the living and inanimate environment, short-term and long-term objectives, et cetera)[12], and we will deal with this category in more detail in the following chapters (see chapter 2.7.).

---

12 Biology defines life as a set of signals and self-sustainning processes in the body of specific organism, which ensure metabolism, sensitiveness and reproduction.

Now we are done with the rough description of how a computer functions and will try how to see a human as an improved robot. May I ask again not to look for exaggeration in such a description as it is the implementation of the same principles and laws of nature to a functionally similar natural item:

In essence, a human is a biological apparatus with the primary objective of "survival" (the continuity of existence and functioning) and everything subordinates to it except the highest level of rational activity that contains the morals and honor (thus we admits a suicide). This apparatus functions as a continuous feedback system in such a way that it acquires and evaluates information from the outside and makes an internal decision either for an action, which would be of an advantage for the apparatus' objectives as defined by a previous generation, or own ongoing automatic activity created during its existence (long-term and strategic) or creates its own objectives (short-term and thus tactical to meet the long-term strategic objectives). This cycle keeps repeating itself, typically many times a second, in line with the supply of updated external and internal information provided by the receptors (sensors) – senses.

I will introduce a more detailed description of this basic principle in the following chapters using notions and tools from computer science, robotics and software engineering. In order to visualise better not only these notions and tools but also the basic functions they perform, I am presenting here the basic charts with building blocks of a computer, a robot and a human so that their similarity is shown, in spite of the fact the fundamentals may be a thousand or million times more sophisticated in a human than in a computer and a robot.

Block chart of a computer:

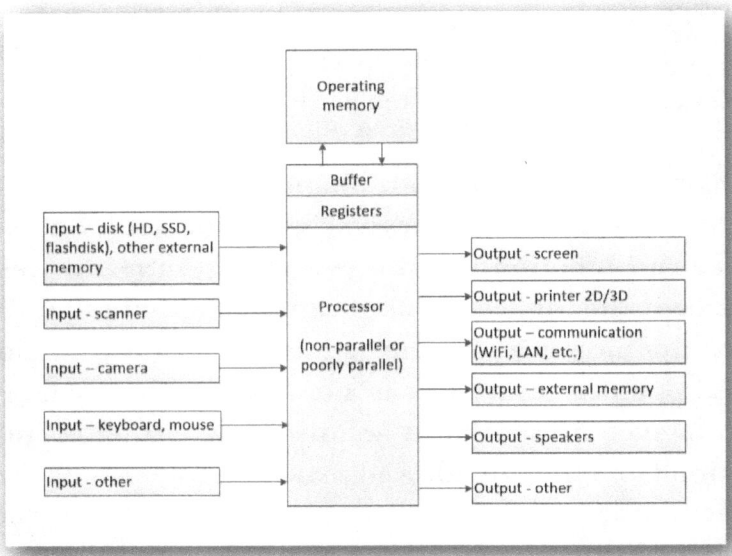

Block chart of a robot:

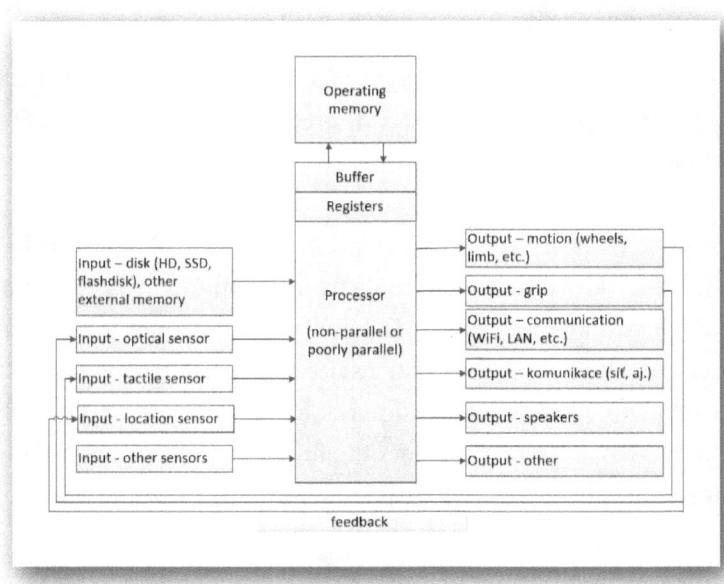

Block chart of a human being:

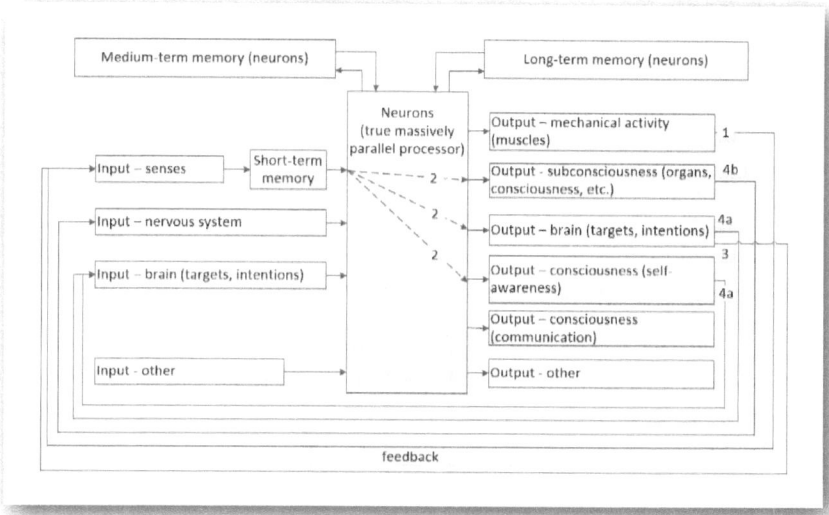

2.6. Operating system

The operating system is an important and necessary component not only of every computer or robot but also of the living organism. Without it any computer or robot or human body would be only a mass doing nothing though it may the most sophisticated product or organism. The operating system (the usual abbreviation is OS) performs fundamental operations on the basic material parts (hardware), for example it performs the command to show the text HELLO[13] on the display (that is, it switches on the electrical signals in the display so that the right light points are activated to show HELLO), or it performs the command to print the text HELLO on a printer (using electrical signals to transfer the toner or the ink on paper so that the text HELLO is permanently shown), or it issues a command to the human arm to move in a way that

---

13 In most programming textbooks, usually the first example is given of a program output (on screen or on printer) as the text "HELLO WORLD". This is the announcement to the world that the program runs.

is precisely defined for this movement. Similarly, the OS commands the recording of an image from the human eye or the robot camera into memory and makes it available for further processing at the OS level or at the level of another higher-level program (the human will or an application program).

In fact the OS of a human or animal may be compared to the unconditioned reflex or the basic functions of a human at the subconscious level – none of us gives a signal to the heart to beat. Some people (yogis and others) can consciously signal their subconsciousness to influence their heart frequency (slow it or even stop it for a while), but this is a very special technique for which such a human being must train for long years. Another example of subconscious activity is food digestion – it is being done completely outside our consciousness.

2.7. Moreover, what about the human soul?

In the above chapters I intentionally mentioned some of the properties and functionalities of an OS (operating system) in such a way that from the computer science viewpoint there is no principal difference in the functionalities of an OS in humans, animals or robots. It is the OS which performs the basic operations on hardware (human or animal body, robot elements or just the printhead). It may again seem to some people as too much of an exaggeration but matter does not do anything[14] on its own account and whatever is non-material and performs some operations on matter is, in fact, a program or software (processor instructions or a thought). The bridge between software and hardware is either on an electrical basis (a very weak signal from the processor is augmented to drive a strong electrical engine of a robotic arm) or on a chemical basis. Processes in the human body are either on an electrical basis – electrical signals in

---

14 I do not mean such invisible processes as when iron corrodes, water evaporates, et cetera

the brain are nowadays quite routinely measured by EEG – or a chemical basis – propagation of stimuli in nerves.

There is a consensus today that human mind (soul) is divided into consciousness and subconsciousness, but sometimes other notions are used in this connection without explanation of what the notions mean from the computer science point of view. From the medical viewpoint, the idea of "consciousness" is understood as contradiction to unconsciousness, sleep or hypnosis. Computer science, however, needs (mathematically) exact definitions and inclusions – nothing may be done in a computer "somehow". It is done in one way or another but never as "something in between" without an additional complete and precise description[15]. Why do I speak so imperatively about it? I am trying to differentiate between the so-called **exact sciences** and so-called **empirical sciences**. The exact sciences always work with precisely defined terms, which are in mutually well defined relations. The empirical sciences, in contrast, have no such precise terms (at least not until now) because so far the subject of research of an empirical science is also not described precisely. I am trying here to use computer science methods and terms to describe those notions and fields that have so far been described only generally in sciences like psychology and psychiatry.

We can describe consciousness as the state of being conscious of oneself – while an animal perceives the world as "me and the environment", a human perceives himself and herself as a part of this "environment"; however, with specific properties and position in this environment. In animals we need to find a name for the entity which differentiates them from humans as animals have no mind based on will, imagination and abstract goal definition. The animals are limited to reflex activities (hunting

---

15 This approach corresponds with "fuzzy logic", which works with values "something in between" (usually between 0 and 1) and the number of such values is infinite. The fuzzy logic works with a degree of truth, i.e. quantification of proximity to the one of the extremes. Such a value "something in between" is concrete and exact.

for food, mating,...) which are performed subconsciously (by reflexes), that is there is no planning of the sort "now I am hunting for some food, finish eating and will visit a colleague in the neighbor territory to discuss how to jointly expand our territories and increase our packs". ☺ Even the protection of cubs by a female, and she is ready to die to protect them, is reflexive (top priority reflex), whatever it can it can looks like an affection or love. It is female reflex "only" for the good of the pack and species generally.

Analogously, we need to distinguish between the states of subconsciousness and unconsciousness which appear to be similar — they differ primarily in the fact that in unconsciousness, consciousness does not work. Unconsciousness is "only" an outage of the functional subsystem of consciousness, due, for example, to a lack of energy, or a blow on the head, et cetera. Moreover, subconsciousness works both in the conscious state and the unconscious state. There is, of course, significant difference between unconsciousness and sleep, which could appear quite similar. The difference is that sleep is a regenerative process, which is triggered according to a plan and contains essential activities for the long-term existence of the particular human, while unconsciousness is the human defense against an unwanted state (depletion of energy, when the rest of energy must be preserved to maintain life functions, or it is necessary to restart consciousness to restore the normal function of it).

In subconsciousness, the basic life functions are maintained (breathing, heart frequency, digestion, et cetera) and sometimes even data acquisition from the environment works (in a coma — grave unconscious state — the basic life functions perform well, and it is proved that even in a coma a human can sense voices from the environment and remember them and process them).

A FEW CRAZY IDEAS

"You will be kind of woozy after those pills but in your case it makes no difference".

Illustration © Pavel Kantorek

From this point of view it is necessary to structure in detail the individual layers of the psyche of creatures or, as the case may be, of the separate layers of software:

	human	animal	robot
consciousness	Higher consciousness – highly abstract thinking activity[16]	Does not exist	Does not exist
	Medium consciousness: abstract thinking activity[17]	Medium mind: mental activity targeted on providing of vital necessity[18]	application software for specific tasks
	Basic consciousness (awareness[19]) – mental activity at the level of the conditioned reflexes[20]	Basic mind (awareness) – mental activity at the level of the conditioned reflexes[21]	Software shell above the operating system
subconsciousness	Subconscious mind (information control with high-level abstraction): higher mental activity independent on consciousness[22], unconditioned reflexes, processing of experiences with creating of rules and relations between them, coding of experiences into the compact form.	Higher mental level of activities with low-level abstraction: (unconditioned reflexes) – recording of what the animal sees, hears and smells into short-term and long-term memories (mechanisms of processing are similar to the human)	Operating system's core – file management, programs starting, et cetera, high level complex functions
	Subconsciousness: basic mental activity – control of body without regard to awareness, sleep or hypnosis (digestion, heart frequency, breathing), low-level senses activity	Basic mental activity – control of body without regard to awareness or sleep (digestion, heart frequency, breathing), low-level senses activity	Basic control software (drivers)
matter	body	body	hardware

---

16 For example, strategy optimization of hunting or similar activity based on the generalization of experience from several branches, an invention, an artwork, etc.

17 Goal definition, the will-based realization or decision, etc.

18 Change of strategy after the animal finds itself in another place – on a savanna instead of a forest, etc. Such a process takes quite a long time and is not the result of an abstract thought.

19 See Glossary.

20 Acquired / learned reactions – commonly known is the experiment of I. P. Pavlov, Russian physiologist, psychologist, physician and Nobel Prize winner: he was looking at salivation in dogs in response to being fed, when he noticed that his dogs would begin to salivate whenever he entered the room, even when he was not bringing them food.

21 Experience acquiring – prick with a spine, instinctive behaviour of a group of animals, etc.

22 Subconsciousness works both independently of consciousness and at the request of consciousness.

2.8. Definition of life

I strongly hope that you will continue reading this book even after reading the next lines. It is courageous to even try to define the concept of "life", even more courageous is the hope that I am successful in approaching close to the truth. I am glad to live in an era when heretics are not burned at the stake. ☺

We can define life as the existence of a feedback system with these feedbacks or reactions –

- "outside to outside" (feedback No. 1 in the chart of a human in chapter 2.4. above) – stimulus comes from outside, and the system also responds to an outside phenomenon or state (responds mechanically but does not learn),
- "outside to inside" (feedback No. 2 in the chart of a human being in chapter 2.4. above) – external stimulus influences internal processes of the system and decision-making, primarily by sending information "inside" for processing and decision-making in order to maintain its existence and the chance to have offspring (that is, it learns from stimuli and events from the outside),
- "inside to outside" (feedback No. 3 in the chart of a human being in chapter 2.4. above) – will / interest to influence the environment mainly with a view to maintain its existence and a chance to have offspring (there is a plan or intention of the system to be achieved which is used to direct the activities towards the external world), and primarily,
- "inside to inside" (feedbacks No. 4a and 4b in the chart of a human being above) – the system is aware of itself, in addition to its awareness of the environment, and

considers itself to be a part of the environment (the environment forms boundary conditions of its activity) including the possibility and tools to influence the environment in accordance with its own interests and objectives (it creates rules for its functioning that are further optimized for the major objectives of the system – organism) (4a). This is an autoregulation process and "self" is a part of the process.

As soon as such a system starts to be aware of itself, it considers itself a part of the environment in its model of functioning and pushes forth its interests (after it created them), we can call it life. The awareness of itself can mean just acting in line with the interests of a group of similar organisms with which it identifies itself, i.e. of appropriate species or genus in biological terminology.

The main objective of such a system is to preserve its existence (as the genus or generation) both in the short-term (provision of necessary resources) and long-term (provision of an offspring owing to a limited life span of every material system). Such a system may rest on a biological or other fundament. As soon as we use an external force to violate or destroy whatever is aware of itself (we kill it), we can call it life (no more alive … ☹).

This definition of life works even for organisms lower than the animals: plants, bacteria, viruses, molds, coral, et cetera. It is just necessary to put some activities into a more precise form. These organisms have no offspring in the proper sense of the word, but they multiply, which is one of the basic life objectives – to preserve the existence. The feedback (the response) "outside to outside" and "inside to outside" does not proceed mechanically (using arms or legs and similar), but it can be on a chemical (releasing the specific chemicals which correspond to the interest of such an organism) or a biological basis.

A very simple definition of life could then be: a thing is lifeless until it starts to claim what it wants, and before it starts to develop those requirements. ☺ Then it will have defined its interests and continues pushing them forth.

If we use this definition of life, we can then define some notions of life using the terminology of the computer science thus:

> **Soul** = an organized feedback computational category (software), which includes the function of consciousness of itself by working with feedback towards itself and from itself. Hence, it is a feedback system not only externally (it responds to a change of state in the environment), but first of all internally (it performs activities in view of its intentions or with intentional impacts on itself). An operating systems **IS NOT** a soul.
> **Life** = the activity of a material system (the body, the robot's frame) with the soul.

The definition states that life is the activity of the soul in the material environment (the body, robot's frame) while the soul itself is the non-material computational category of the software type. I can store the information "zero" or "one" to a material carrier, for example, by magnetizing a part of the material (on the surface of the magnetic disk), whereby the information "zero" is represented by the magnetic orientation "north-south" and the information "one" is represented by the magnetic orientation "south-north". However, this material storage does not change the fact that the information exists even without storing it into the material carrier. Software could be written to the main memory (as a runable instruction, i.e. the one which runs as the program, as opposed to the data which "do nothing" but waits for processing), the identical piece of software could be written on the flash-disk or another external memory (as a file) or I can print it on paper (special type of file) and it is always the same program which may or may not be connected with some material (carrier). If we, however, view this software as a mathematical organized structure, then the program need not be recorded nor printed anywhere and yet still it exists. Only its linkage with material gives it the possibility to execute its functions and become evident to our senses. Software could also be considered as a region with lower entropy.

One point of view how to distinguish living structure (organism) from non-living structure (machine) is the Third Law of Robotics of Isaac Asimov[23]: **"A robot must protect its existence as long as such protection does not conflict with the First or Second Law of Robotics"**. Even a robot which obeys all Laws of Robotics[24] is not a living creature, the reason being that it has no interests of its own. Its purpose is to serve, as its own interests are not defined. A robot could become a cyborg or a living creature only by creating its own interests and starting applying them.

If the Third Law of Robotics is applied, the internal feedback is done only to retain system activities as a functional system and not to push itself forth. If it were starting to push itself forth, it would not simply

---

23 **Isaac Asimov (Izak Judovich Ozimov**, January 2nd, 1920, Petrovich, Russia – April 6th, 1992, New York, U.S.A.) was an American writer and biochemist. He published over 300 scientific, popular science and sci-fi books. He has got many prestigious appreciations. He is the author of three Laws of Robotics, see below.

24 Isaac Asimov's Laws of Robotics:

1. A robot may not injure a human being or, through inaction, allow a human being to come to harm.
2. A robot must obey the orders given it by human beings except where such orders would conflict with the First Law.
3. A robot must protect its own existence as long as such protection does not conflict with the First or Second Laws.

Later on the zeroeth law of robotics was created:

0. A robot may not harm humanity, or, by inaction, allow humanity to come to harm,

together with adding the followinf text to the other laws of robotics:
"except where such orders would conflict with the Zeroeth Law"

serve but strive to maintain and improve functioning of itself as if it were a living system (life ☺) and to create descendants. The descendants are a new generation of the living system which are considered as a continuity of its existence and on which it may pass its experience acquired during its existence and, logically, experience taken over from its predecessors (parents).

I find it very important that the soul as an organized functional structure reminding us of software may exist independently of matter – it may be in the design phase (of this software) only and lay dormant on a data carrier (as the software was not yet installed into the operating environment and did not start to run). It not important for the existence of software whether or not it is stored on a material carrier – it may exist without being stored, i.e. in the abstract form (just logical form, not stored). This is a view where many people strictly disagree with me.

I put all this in by intention, to foster discussion whether or not there can exist a soul without a body. From this (my) point of view, the answer is unambiguous: YES, there can exist a soul without a body, for example in the form of an immaterial mathematical or computer science structure. The fully legitimate question related to this assertion is: how would the soul be recorded (on which carrier or into which other operational environment it would go) after leaving the body (its operational environment)? But this is the technological issue.

We may lighten the seriousness of the quest for the definition of life using the following joke:

Do you know what life is?
Life is a mortal disease transmitted
by sexual intercourse. ☺

CHAPTER 3

Do you belong among the "ten percent" people? Be happy!

3.1. How well do we use brain capacity?

We often hear that humans use only ten percent of brain capacity[25] and how stupendous it will be when we learn to use the rest ninety percent. Then it will be a brave new world when all the people will be (roughly) ten times cleverer ☺. The world will be full of intelligent people and everything will be beautiful. I believe this is one of the terrible fallacies and simplifications we commit, or better said were committed by experts who had the feeling that such was the reality (the opinion does not appear so often nowadays). I will now explain the reasons.

It is not clear how people arrived at the figure 10 percent of usable brain capacity. Why not one percent or fifty percent? In every case it is an estimate based on the assumption that we do not use the full brain capacity in our rational mode (consciousness). It is not important for this chapter whether the number is one percent, ten percent or fifty percent – we use the 10 percent as our working number (such a number is the usual one). It appears almost certain to be true that in the rational mode (consciousness)

---

[25] For example see http://www.cyberspacei.com/englishwiz/library/friends/how_to_win_friends.htm - Lowell Thomas's foreword to the Dale Carnegie's book "How to Win Friends and Influence People".

we use only a part of brain capacity, but we deeply underestimate what is going on in the other part – it does not, really, goof off or skive off.

Humans came into being by a specific evolution of the animal – by development of brain activity, specifically by a formation of ability to abstract. Just by that brain evolution we became the creatures who, in addition to animal reflexes (yes indeed, we are just little animals overcome by various instincts), has a certain superstructure (consciousness) above the lower layer (subconsciousness). The superstructure (consciousness) works on the top of the lower one (subconsciousness), which is an operating environment for the upper one. The consciousness works as a special task in that operating environment (subconsciousness). The consciousness provides for rational thinking including demanding tasks, such as the ability of abstraction. I will go into more details later in this chapter. Just to elucidate what kind of nonsense may appear in comparing a computer to humans, I introduce one more example.

Comparison of human brain with computers is really like skating on thin ice. It is because we humans hold computers in awe (after all, they are imposing machines; however, it still holds good that they are only "fast idiots" doing simple things very fast). If someone, who may understand what goes on inside a computer (this is the better case) but does not at the same time understand the complexity of tasks tackled by the human brain, now tries to compare humans with computers, we can have a myth, deeply rooted because "computer experts said so". One such a a fallacy appeared strongly in the eighties of the last century and was presented also in the nineties. Technological development revealed more and more of its absurdities and this is good as the essence of such a comparison was clearly seen.

As soon as I started to engage with computers (cca in 1980) I kept hearing a similarly big and terrible fallacy as the one introduced above in this chapter (namely, that humans use 10% of the brain capacity only) which in addition looked rather credible. After all, it was pronounced by the experts, and the laymen (ordinary people) liked to believe we were striding towards the creation of an artificial brain and would soon be able to create one (please understand: no need to learn, someone will store it all into my brain and I will become a genius). In this case, the cause was a confusion in the notions of memory capacity, computing performance and architecture.

That logic about the comparison of a human and computer went on like this:

A human has about 100 billion neurons, that is a specialized nerve cells in the brain. Computers in 1980 or thereabout had the main memory of roughly hundreds of thousands bytes (byte is a memory cell of eight bits and can store one number between 0 and 255) and external memories had capacities in millions of bytes (megabytes). The fundamental fallacy was that a single neuron in the brain was seen to be as one byte; however, I did not catch a single article then which would sweep this sheer nonsense from the table straight away.

I have to say that I saw the absurdity of this logic immediately on spite of having rather little experience at that time (even now I see myself not as an expert, nevertheless I know a little more about computers than at that time). During that fantastic racing competition of mankind and nature, everybody expected that humans would create a device of a 100 billion of something (at least bytes) so that it could compare to the human brain. As the capabilities of computers grew, so did the desired goal get closer – in about 1990 the computers had millions bytes (megabytes) of main memory and hundreds megabytes of external memory, about 2000 as many as hundreds megabytes of main memory and tens gigabytes of external memory. Tens gigabytes stand for tens billion of memory cells and this was "in the offing" of a brain with its 100 billion neurons.

As technology of memories progressed (main memory, disk memory and other external memories), the myth started to disappear because everybody felt "there must be something wrong in the logic". Well yes, for sure, when we compare just one single memory cell containing eight digits, zeroes and ones (one byte), with a neuron which is a supercomputer even from nowaday's scale view (2016) whose functionality and efficiency we cannot even estimate, much less build. I did the estimate of the performance of this apparatus — a single neuron cell in the human's brain (or elsewhere in the nervous system of human body) — in chapter 2.3. of this book (Supercomputer genius in our eye). However, this estimate is made from a position of great humility as the human being is a much more wonderful machine than we can yet imagine. As our imagination fails us for such an unusual take on the situation, such an experiment requires both humility and great courage.

A similar absurdity as the above given comparison of neurons and computer memory capacity is the **total transfer speed of the connections of all neurons in the human brain** (that is, interconnections between neurons — synapses). Even recently (2012), non-IT experts — such as neurologists[26] — give this number roughly **one terabit per second**. Sounds good, does it not? Such a tremendous performance we have in our head? Well then I am not so stupid as some may think. But this is another nonsense. Why? Let us calculate what it means: Further on, I state in chapter 3.3. (based on the human brain specialists) that one neuron has about 1,000 interconnections (synapses) to neigbouring neurons[27]. Simple calculation gives the result as 500 trillion synapsis. Some sources give the number of synapsis in the brain just 500 trillion[28] (5×10^{14}), some other sources give 10^{16} synapses[29], so let us make an estimate of 10^{15} synapses. The speed one terabite per second, which is the guess of brain-specialists

---

26 source: MUDr. Frantisek Koukolik: Social brain: several evolution principles (2012), see https://www.youtube.com/watch?v=PBkC2XxzxeY

27 the science source quote generaly the number of synapsis between neurons in the range of 100 and 20,000. Those number will be probably right, altough the range is large, because various tasks need various type of communication connection between logic / computational units (processors). I will work here with mean number 1,000 synapsis per neuron.

28 source: web pages of MUDr. Zbynek Mlcoch: http://www.zbynekmlcoch.cz/informace/medicina/nemoci-lecba/cisla-v-medicine-zrak-sluch-kosti-svaly-mozek-kuze-traveni-vylucovani

29 source: Academic bulletin – official journal of Czech Academy of Science, see: http://abicko.avcr.cz/2013/07/06/vyzkum.html

of the speed of communication between neurons via synapses, seems really a fast one, but this is "only" 10^{12} bits per second (ones or zeroes per second). When we divide the speed 10^{12} bits per second by the number 10^{15} synapses, the speed we get is such that one synapsiscan carry a one or zero in 1,000 seconds (20 minutes). We would be long dead (by an attack of an animal or opponent). Even if we assume a rough and completely sober estimate that one synapsiscan transmit 100 bits (ones and zeroes) per second, we have a nice 10^{17} bits per second per synapse, which is "only" 100,000 times more than what is given among the experts. ☺ In reality, the synapses can transfer much more than just 100 bits per second during its complex activities. Something is rotten in the state of Denmark[30] ... So far, we have not come any closer to understand the human brain activity.

It is beyond belief how often similar mindsets and forecasts, based only on – "how beautiful it would be" – appear even today. At the time this book was in its final stages, another news like the ones above appeared, and at that it is an activity by a very renowned person who founded an university for emerging technologies. This renowned and reputable person[31] is behind quite a number of key technological inventions of the past forty years and he is a remarkable personality. His co-workers at the university forecast however, that "in the near future, computers will have the capacity of the human brain" (though so far nobody knows what this capacity is – we do know only that the brain has a 100 billion neurons and 500 trillion synapses but no capacity follows from this), then "technologies will spring up which will overcome human intelligence" (so far however, no one knows, what is the basis of human intelligence, though we plan to outperform it in 25 years), then that "brain implants will improve human memory", (no one knows until now how human memory operates, and in addition human memory is being enhanced even today by the Internet and search engines like Google® which find whatever we do not remem-

---

30 citation of William Shakespeare (1564 – 1616)
31 I do not name knowingly the name of the person and the university.

ber). Perhaps they have in mind a small but a full-fledged Google®-similar systém (or at least a direct link to it), or then also that "human intelligence can be increased by injecting millions of nanotechnological devices into the organism, each the size of a blood cell". So far however no one knows exactly what human intelligence is but we can improve it by injecting devices which may be taken as hostile by the human immune system and it will try to destroy it, moreover we know nothing about the fundamental principles of the immune system – distinguishing among "friends or foes" – as well ....

"This is the brain, my boy. It does weird things to us humans".

Illustration © Pavel Kantorek

I may know what marketing is but we need to be down-to-earth and humble to nature. I esteem highly this person as well as the activities of his co-workers at the university but the above forecasts are quoted almost literally. One thing is to improve human senses (for example, a 100,000 times more sensitive smell to get us to the level of dogs ☺), or adding the external sources of information (after all, books have been doing this for centuries), another thing is to talk seriously about the improvement in brain function and intelligence when so far we have no idea about what it is and how it works (not even in principle). For example, we have at least some feeling how a new cell is created using DNA, we can influence it (using the "trial-error" method), but are we able to improve it? By far not. We do not even know what is wrong with it so that we can improve it ... ☺ We have no choice but to wait these mentioned 25 years and keep our fingers crossed for the scientists (not only from the stated university) or correct the forecasts along the way.

3.2. Subconsciousness never sleeps

The devil also never sleeps allegedly but here we have something different – subconsciousness is a (software) part of our body. According to a widely-spread opinion (see above) our consciousness is using only 10 percent of our brain capacity and the rest is unused. It is quite all right to determine that our brain uses 10 percent of its capacity for some activity; this, however, does not mean that the other 90 percent is idle. Quite the contrary, this 90 percent is probably used by our subconsciousness that is more important when you compare the importance of consciousness and subconsciousness. From the viewpoint of brain architecture, consciousness is probably only a superstructure on subconsciousness. We can even use an IT comparison – consciousness is a specialized processor for a certain task. This specialized processor (consciousness as the functional level of human rational and abstract thinking) is only an attempt of nature to find a more efficient form of survival in the evolution of species. We will see if this is the right way or if this is an impasse. ☺ Do not be scared – it will be some long time before humans learn it and our generation will not be there.

In principle, we can understand subconsciousness as the basic operating environment of a human being (it has been so since the times of humans as animals) which runs the task "consciousness" using for its operation all the resources of subconsciousness (memory, performance and access to I/O devices – muscles, senses, and other) wherever these are required by the task "consciousness". Consciousness, from the viewpoint of the computational brain architecture, may be considered to be a specialized processor for the core computational part of the human brain – subconsciousness (what the humans had before they started to change from apes to Homo Sapiens). Such a specialized processor, capable of abstract operations, would then serve primarily to secure a better way of survival than what could be provided by processor "without imagination", that is subconsciousness). Pleasures (the perks of life) could then be only the secondary functions of such a processor. You may not like this. We live to enjoy life, don't we? ☺ However, nature does not allow any meddling.

In this mindset, we can understand schizophrenia (split of mental functions) as a software error in the task of "consciousness" (and perhaps an error of the operating environment – subconsciousness – which allows this to happen), when two instances of an application "consciousness" switch over one to the other, or even more such "consciousness", which however do both work with the same background subconsciousness data – I mean, for example, the abilities, experiences and medium-term and long-term memory. Thus it is possible that a schizophrenic person, when it switches over to another application of "consciousness" (to the Human No. 2 – the human behaves as "someone else"), has at the disposal, at least partially, the knowledge and experience of the application of "consciousness" (Human No. 1), as these are stored in the subconsciousness.

3.3. More heads know more

Not only that more heads do know more but if we understand "head" to be one neuron (which knows a lot), then many neurons will know much more. The issue is not only that one neuron connects to a colleague neuron in the neighborhood (or two, three, ... colleagues) but mainly with how many it is connected (direct connection – if each neuron is connected to at least two of its colleagues, then all the neurons are interconnected, albeit indirectly). A much higher level of "cooperation" happens when each neuron is connected with more colleagues than only two. According to modern science one neuron can be connected directly with 20 to 20,000 "colleagues" (by various sources, some claim "only" 20 to 1,000 connections, other sources claim just 20,000 connections) – let us take the average of 1,000 connections. Such a connection of two neurons (or a neuron and receptor cell – a sensor) to carry impulses is called a **synapsis**(imagine that as the cable).

We can certainly imagine how adroitly the neurons operate when each of them can request help from a 1,000 colleagues (at the same time!) about a thousand times per second, and each of those can request the same from a similar number of colleagues, so that in two or three of these steps we may have millions or up to billions of neurons at play. We can only dream about such (truly parallel) processing even in our existing computers (at the processor level, I mean). The only devices coming closer to this concept are computer networks of companies like Google® or Facebook®, which have, as a minimum, hundreds of thousands or possibly millions of computers organized in this way. However, let us remember again the comparison with the human retina cell in chapter 2.3. of this book (Supercomputer genius in our eye) when all of the existing computers (even a supercomputer) may pale in envy in front of even a single neuron. Such a processing is called massively parallel which means that a large number of computers or processors cooperate on solution of one task. The issue here is not only the number or the way the computers are interconnected but the much more important thing of how the task is divided into many parts so that parallel processing may have some sense.

We sometimes see today that the "unit of activity" of the human brain is just the synapsis rather than the neuron. It is the same fallacy as if we were saying that today's computer science is the Internet (that is, the communication network) and that the quality and performance of a computational system depends on how well they are interconnected only. And we would not take into account the importance of the activity of computers with the tasks running on them, the performance of their processors, the number of these processors, number and speed of storages, and similar considerations. I believe that all of us can feel the absurdity of this view, even without being experts on computers or brains. A racing car is not only the tires or transmission only though they may be the best ones in the world. ☺ What use would an excellent communication network be (with servers used exclusively for computer communication) without those computers that talk to each other through this network (the Internet) and work on a specific task? Such a network would be nothing but wires. In a similar vein, what use would a million or billion computers be, were they not able to "talk" each other and cooperate on tasks, including a specialization on a particular task. YouTube® service (an estimate is hundreds thousand to million servers) is such a special task - storing movies. We can take YouTube® as one logical computer implemented by a million "boxes", names we sometimes use for the physical computers proper.

Of course, computers as such (without communication with other computers) have practical importance, but are limited in comparison to computers that are interconnected and communicate each other. After all, before worldwide networks of Internet type came into being, computers were fully used, often interconnected by local network (tens of computers typically) only. Just a few computers (even as few as two) can provide a completely new logic in information processing (for example, the client – server architecture when one computer requests something from the other one that is specialized for such an inquiry). This architecture originated at the end of the nineties of the last century when processors were not as powerful as nowadays. The present powerful processor is capable of dividing the power into several logically independent virtual computers all running on one physical machine[32] but behaving independently – it is called virtualization, and the fictitious independent computers are called virtual machines (computers). Here the case is not so much for a new logic (this is identical with the logic in older NON-virtual and with the modern virtual computers). The issue is the efficient use of the power of the modern computer – why should we buy three computers for three specialized tasks (three boxes) if the same will run on one powerful computer (one box) which pretends to be three computers?

Let us, however, go back to our understanding of neurons and synapses as one whole rather than preferring the synapses to neurons (or vice versa). The neuron is nature's work of a genius. Same as a synapse, the mechanism of communication among neurons is the work of a genius. However, the roles of these elements are not separable, same as the roles of the computers and the Internet (communication network). Only the

---

[32] For simplification, I do not consider that even today's ordinary computers (PCs, notebooks, et cetera) has multi-kernel processors, where kernel is an independent processor from the historical viewpoint of 10-20 years ago. Today's ordinary servers have several such multikernel processors (usually 2, 4 or more). In all above-mentioned cases, the processors (or processor kernels) do not know how to interact and cooperate as perfectly as neurons do – some types of neurons (specialized very clever neurons) are interconnected by 10,000+ synapses with neighboring neurons, i.e. direct communication channels. If even one such a neuron is interconnected with 20,000 neighbor neurons via 20,000 synapses and the communication is parallel (of course, each of the 20,000 neurons are further interconnected with other neighbor neurons), then parallelism is so high, that even today's massive-parallel computers ☺ are just primitive machines comparing to the neuron and synapse solution in the brain. Those nowadays massively-parallel computers are massively-parallel just on the data level – for example every of, let say, one million processors, which are interconnected quite poorly, has a small part of data to process. After finishing the task, the result is sent to the main processor ("chief") which puts the million parts of data together to create the aggregate result. This way of computing is used for weather forecasting or in the graphical adapters in the "game-PCs". Key limitation here is the "bus", which is the interconnection channel between logical elements (processor, memory, sensors, et cetera), and the synapse is such a bus. Probably not a single nowadays computer has 20,000 buses working fully and cleverly parallel in the same way as neurons do (I am not sure if there exist a very special supercomputer with such a number of buses but, if so, it is just one computer comparing to 100 billion neurons). The brain is a "computer" with 100 billion of very clever processors / computers working really parallel and communicating via 500 trillions of buses in parallel. The human being is really an incredible machine. ☺

A FEW CRAZY IDEAS

interconnections of neurons (about 100 billion) and synapses (about 500 trillion) and furthermore the organization of their working (we are just at the very beginning of understanding here), is the quality representing the brain in all its beauty.

What is important for this massively parallel processing of tasks is the organization of work (architecture of the application software) among the computing units (computers or neurons), which substantially influences the efficiency of the task computation. The performance of a billion terabit-per-second LANs, for example, can be devastated by a silly design of the organization of task processing (inefficient or faulty). A silly design means, in this case, for example, to look up a single definite item in a database of a billion data items by moving the whole billion data items to the processor on quite slow line one at a time, where 999,999,999 are discarded and the last one is finally the one sought (if we have bad luck, and the sought item is the last one). In this silly solution, we completely congested the communication channel (network) by the 999,999,999 data items as the network is always slower than the information processing speed of the computer. A clever solution (used normally today) is when the computer sends a request (the specification of the sought item) to the location with the billion data (storage server) where the requested item is located much faster and is sent back (one item against a billion items) so that rather than sending a billion data items, the item is sent only twice (request there, answer back). Efficiency is clear to everybody at the first sight. It requires, however, some thinking, and as a classic scholar once stated, thinking is painful. ☺

Another clever speedup of tasks is the method (process) used in looking up a data item, as we mentioned a few lines above, directly in "the location of the billion data items." It is well known that searching for one data item in a database (set of data) may be item by item (sequential), which is very time-consuming and slows down with the growing size of the database (by orders of magnitude – a thousand times slower, million times, billion (10^9) times, 10^{15} times,...). Another type of search, other than the sequential search, is "indexed search". We sort the data before the search (this is done only once at the beginning) and then search in one half of the database, then in the half of a half and so on (so-called halving of the interval) so that instead of a billion steps (10^9 steps), roughly 30 steps are sufficient by halving the interval to locate the requested item (instead of a 10^{15} steps we need only about 50 steps of bisection).

As we can see, the same task can be done using the clever method in the unit time "1" and using the silly method in the time of "1,000,000,000,000,000,000" and this can go up to tens of years even with a computer processing 1,000,000,000 instructions per second (one billion seconds is about 32 years). ☺

Another case of mistake in today's[33] understanding of the human brain elements (neurons, synapses and processes running on them) is that it does not differentiate between hardware and software[34] in the case of the human brain. This understanding claims that it is not possible to distinguish hardware and software of brain and that it is just a single entity while with computers we can easily separate hardware from software. What is the idea behind that statement? This theory states that brain changes in the process of its activity by creating synapses; for example, when we listen to someone's speech, the brain creates synapses as it listens and understands a sentence[35]. It is true that the brain changes while listening and understanding a sentence but no new synapse or axon is created, only information (signal) propagates through them, i.e. the synapses are activated (used for transmission of information) and/or reconfigured only. The creation (growth) of new axon ("wire") takes a much longer time (several minutes at least) than reading and understanding a sentence. In essence, the mat-

---

[33] Year 2012 – it is "today" from the global view
[34] The source: MUDr. Frantisek Koukolik: Social brain: several evolution principles (2012), see https://www.youtube.com/watch?v=PBkC2XxzxeY
[35] The source: MUDr. Frantisek Koukolik: Social brain: several evolution principles (2012), see https://www.youtube.com/watch?v=PBkC2XxzxeY

ter is reorganized, that is, the matter is not changed, "only" its settings are changed. The creation of a new synapse in the way of growth of new axon / axons used for the communication between two neurons is the consequent optimization of brain mass (hardware) organization for the next case of listening to someone's speech or similar brain activity. I assume that it is just the misconception of hardware and software terms.

Synapse is a general term for the connection of two neurons, and the material connection is in the form of a so-called axon, kind of a biological wire. This wire can be rather short, only to the next neuron, however, the longest axon measures about two meters and connects a neuron and some places in the toes of a foot. As we stated above, new synapses for the direct connection of two neurons are not created when we listen to a sentence as the time for the creation (growth) of a new synapsis(a new axon must come into being) is roughly a few minutes at least. In the course of listening to the sentence, synapses "only" activate (pass on signals – we call the signals nerve impulses or also action potential of a neuron) and this signal either wears off or can be stored somewhere (we do not yet know where and how) as a memory item. The time of execution of such a nerve impulse is about a thousandth of a second which means that one neuron can send about a 1,000 nerve impulses per second along one synapse.

From the computational point of view, we do not know at present what is the content of one such nerve impulse, that is which information it sends. I mean whether it is only yes/no, one bit, or whether the information in the nerve impulse is more complex however it is encoded. The fact is that chemical substances called neurotransmitters are responsible for sending the impulses. There are at present several tens of chemical substances acting as the known neurotransmitters (to name a few: glutamate, serotonin, adrenaline, dopamine). The content of the message could then be defined by the type of neurotransmitter or their combination. It can be just the type of neurotransmitter or their combination which could be a message (command) including data (parameters of the command).

In modern computers, we have been using for the past fifteen years so-called XML data, which may contain both data and an instruction for their processing.

It may be quite an important fact that "a single (specific) communication" among neurons is finished only after several nerve signals (via several axons) are passed on, or several nerve signals pass through one axon (if they form one information word by putting together them, so they create the sort of a biological byte). Computer experts, especially those at the lower levels of the network reference model ISO OSI, must certainly laugh up their sleeves, as this is a common process in network communication for them.

And in order to increase even more the similarity to computers, the biologists themselves state that there are three types of synapses (communication among neurons) and namely (careful, a jawbreaker) the axodendritic synapsis (an axon of the first neuron connects to the second one via some connector – dendrite), axoaxonic synapsis (an axon of the first neuron connects directly to the axon of the second neuron, a more direct and faster connection without the connector delay, sort of direct memory access, DMA in computer terminology), and axosomatic synapsis (probably the most amazing one – an axon is connected directly to the cell body of the second neuron, we can say a direct connection to "registers of the second processor", nothing like it exists in computers). Computer experts may feel the chills up their spines. The similarity is evidently not accidental. The issue is that we should not consider neurons and synapses to be something supernatural and beyond understanding but apply to them the laws of computer science. The fact that there run processes inside the neuron so far beyond our understanding (we have no hint even about the principles), does not change anything.

Storing new information onto a magnetic disk does not change the matter (nothing is engraved, nor anything new is created), it changes only its magnetic orientation. The disk surface has either N-S (north-south) or

S-N orientation. This interprets as one bit (zero or one). It is, of course, different for a CD-R disk where one bit is either indelibly burned (the matter changes) into a certain place, or not burned, and the laser beam reading the information in either reflects or does not reflect (states zero or one). This way the information is retrieved in the form of burned or not burned points distanced from one another (so we can see if they are burned or unburned). However, here we only have a special type of memory which does not represent a typical solution.

It is important to note that the same information can have different forms – same as that the information can be stored in some material form (a changed matter, for example, a burned point, or magnetically oriented matter, and other), however, we always have only a tangible record of the information as information itself is not tangible. It is an entity that can be made tangible by recording in a certain way (see above). It is mathematical (computational) category and therefore not matter[36]. That is why we can always distinguish matter (hardware) from information (software). In the case of information, we may, however, distinguish between a data item and instructions for data processing (program).

A note at the end of this chapter: we cannot rule out that the brain functions as an analog computer or a combination of an analog-digital computer (see chapter 2.2). The digital part would then likely be the whole consciousness and upper layer of subconsciousness (see the scheme in the chapter 2.7) to store and sort phenomena (lists of elements, their relations, and other), for abstract mathematics-like operations, predicational logic, and other). The analog approach could then be used in some operations when information is acquired from the environment; that is where the senses record data for the brain.

---

36 see the chapter 2.7

CHAPTER 4

Our good friend Differential Equation or we have our heads full of these

"Professor, I know that you are highly esteemed, but I really don't envy that mess in your head".

Illustration © Pavel Kantorek

A FEW CRAZY IDEAS

4.1. Do you like proverbs?

We can probably concur that the ability to abstract, that is working with abstract notions, plays a key role for any given human being today. The lowest level of the ability to abstract is if an individual is able at least to understand these notions and relations of them (it is the minimum level of the ability to abstract). The higher level of the ability to abstract is if an individual is able to apply that (already understood) abstract model, i.e. to create similar abstract notions based on given abstract model – altough he or she had not created the abstract model, as it was created by someone else, but that he or she is capable using it in similar way. The highest level of the ability to abstract is if he or she can further develop the given abstract model to higher level of abstraction or to generalize several such an abstract models, i.e. to combine several such an abstraction models into one.

By the way, let us note that this division into levels corresponds in humans almost exactly with the elementary education, high-school education (with a school-leaving certificate) and university education. Even higher in the capabilities of abstraction are people like scientists, designer, philosophers, poets and similar others – all these are an extreme example of the ability to make abstractions and no wonder they are often not understood by "normal" people. ☺ In the folk parlance, such a normal human being "is not quick in the uptake".

There is, moreover, a simple way to find the level of abstract thinking for a given human being – the ability to understand jokes or proverbs of various complexities. **Jokes and proverbs are a concentrated abstraction**. I think that researching these entities is a way to understand the human mind. Especially the proverbs are important – see below.

The levels mentioned above of the human abilities to make abstractions as related to the level of completed school education is only illustrative. We have known in modern history (for a few hundred years) several extraordinarily capable people who, because of various reasons, did not even complete an elementary school — one of them is the famous mathematician George Green[37]. With some tongue in cheek, we can say that he was able to conceive quite new mathematical and physical ideas as he was not mathematically "spoiled" by the schools of his era. ☺ However, such cases are extraordinarily rare and cannot be generalized.

The abstraction ability could have been developed in humans in connection with the increasing power of the brain. The increased power of the brain is not directly proportional to its size (it is documented that there are people with a record size of the brain, about 2,000 grams, but are mentally retard). It may be probably linked to the computational architecture of brain function (the way the brain cells are organized and their interconnections) and further to the amount of connections (synapses) among brain cells (neurons). What I have in mind by the above mentioned computational architecture of brain function are the functions not possessed by the animal brain, such as **creating a model of an objective** (intention of the individual) and the related **search for a strategy how to achieve this objective**. There can be many such strategies proposed by a human brain (produced by the volition and efforts of humans — not just of its accord ☺) and the individual in question selects the optimal choice of attaining the objective according to some criterion. This brain feature is implemented by consciousness in the course of searching for new strategies of implementing the same task.

This activity differs from the one of animals who have an objective, too, but it is limited to the basic instincts, first of all to food provision and breeding, which have been repeated in more or less unchanged patterns

---

[37] **George Green** (1793 – 1841) – a British mathematician and physicist, the author of "Green proposition" and many others mathematical and physical ideas. He spent just one year at school in his youth (in the ages 8 to 9) so otherwise he was mathematical auto-didact. He was the son of a miller and he had to help his father in the windmill. He started to study at Cambridge University only when he was 40 (he died when he was 48). Before his studies there, he published important mathematical and physical books, for example the "Essay on using of mathematical analysis on the theory of electricity and magnetism" – the edition was 51 copies altogether. ☺ He finished the university in 4 years with many successes during the study. His study was enabled by his sponzor the mathematician Edward Bromhead, who bought one copy of the above-mentioned book.

for generations. An animal can develop (change) instinct processes only in a limited form and slowly, unlike a human, who may use the willpower and the ability to abstract to find completely different processes to attain the same or better results. Consciousness, from this point of view, is, in fact, a sort of "an abstract engine" for subconsciousness. In distinction to the animals, a human being defines highly abstract objectives (social), such as respect, life objectives, material provisions (which will materialize after the objective is reached), and other, which are all a part of the social life of humans. The definitions of these objectives (determination, design) is a very complicated mathematical and computer science task requiring corresponding computing facilities, i.e. the part of the brain not possessed by animals and which represent a very complex tool. The tool must contain both data (experience) and procedures (algorithms, programs), and additionally also the methods to optimize these procedures and to look for new ones. It is quite the ability to **find new (alternative) procedures** which constitutes one of the extraordinary human capabilities which differentiates humans from animals.

The abilities of the human brain to define an abstract objective originated probably independently of the origin of language (speech first, writing later). Language serves as a basis to form and capture ideas (processes in the brain) for external use so that they may be understandable for other people. I understand the human language as a higher level of communication comparison to animals, as animals need the way of communication primarily for instinct activities. A human language is much more complex as it includes the ability to describe the abstract elements. It includes the ability to express and record (!) the cooperation of functionally independent units (humans ☺) on the interest level, restrictive level (social) and, above all, science level including the tools of king of sciences – mathematics.

Just the possibility to record formalized ideas is exceptional – the whole of nature works in "on-line" mode, i.e. it reacts on the event which exists in this moment, but recording of ideas and methods by human being enables the "off-line" mode, i.e. any human being can load (read and understand) into his or her brain the experience or knowledge of other human being which maybe does not exist at all, and the human being is able to resume the ideas of his or her predecessor and, eventually, he or she can enhance it.

As the human language with above-mentioned features is used for the communication of human beings among themselves, and one human being can be seen as one independent information system (an analogy of a computer or a cybernetic organism), so a language which transmits and passes on the ideas among human beings, could in principle be understood as an inter-human network communication protocol with the feature to store (archive) messages (in computer science, we also use the notion MOM — Message Oriented Middleware, that is the software facilitating the communication between two systems using messages with a given structure).

The mentioned feature "to store messages" is significant as the message may have to wait until it has a chance to be processed (or even until the recipient can understand the message). This is in difference to messages transferred on-line, that is "bad luck to those who are not listening right now". This design is extraordinarily clever and gives stability to such communication and further the resistance against a paucity of sources; that is, if I have no possibility (time, capacity) right now to listen to the message and process it or even remember it, I have a chance to return to it and process it later. Language, the basic means of communication among people, has then both a direct form (voice, of a fleeting character) and an archiving form (writing that survives — cases when the paper burns and destroys the message are not considered).

A language may contain both information (description of reality) and a method (procedure) how to process the information. These methods could be fixed, static (they do not include the possibility of a deviation or variety in understanding the given procedure) or variable — abstract, in the sense that the method must be "inserted" in concrete conditions for a concrete procedure. In humans, these abstract methods of information processing are quite normally recorded using **proverbs**. I believe that **proverbs are the keys to understanding the computational workings of a human in its complexity and the current state of perfection.**

A proverb is, in essence, a decription of relation between elements which the proverb refers to[38] – these are the **rules** for those relations. From the mathematical viewpoint, the proverb is an alternative record of differential equation. The proverb contains in itself a description of an infinite number of similar situations, which can happen, but it only start to make sense then, when we insert it to the concrete situation, i.e. to the concrete **environment**[39]. A specific environment is an alternative record of initial and boundary conditions, which mathematicians use for differential equations solution. Application of a proverb in a given situation, in the mathematical view, is to find a solution of such a differential equation under the given initial and boundary conditions.

I am afraid that I may have annoyed the non-technically oriented readers by using terms "differential equation" and "initial and boundary condition" that they hoped they would never again encounter in life after having had somehow managed them at school (valid for me as well), or which they never encountered. This is the reason I just "fired from the hip" this sentence to make it quite clear where I aim to go and quickly ran away so that a majority of readers may continue reading. ☺ I promise that I will use these (and similar) concepts only so that they may be maximally understandable, and I will always introduce a term that is more or less understandable to everyone and which corresponds to the "nasty" one. I have in mind, for example, that the concept "proverb" which (hopefully)

---

[38] In the mathematical view, it is the differential equation, which describes the relations between elements

[39] In the mathematical view, it is the initial and boundary conditions of a differential equation, which we apply on given differencial equation and we found the solution for this initial and boundary condition, i.e. for specific situation.

does not annoy a normal reader, corresponds in fact to a form of a differential equation. In the following text, I will use the "**rules**" word instead of "differencial equation" and "**environment**" word instead of "initial and boundary condition". I believe that the book will be more readable for most of readers[40].

Most of the proverbs, if you read it as a messages (statements, information), do not make much sense or its sense is so simple that everybody can wonder why something so stupid is presented at all. Some proverbs, read as a message, may though give some sense; however, this message has usually no reasonable value. It is only after we "switch" our thinking to perceive this message as a generalized experience (we make an abstraction), we are opening a space for the meaning contained in the proverb to be understood in a figurative sense.

Let us take, for example, the proverb "can't see the forest for the trees". This message is utter nonsense — if we can see the trees, we can see the forest and vice versa. This is logical. Except that the true meaning of the sentence appears after its generalization, when (1) it is an operation on sets (elements of the set "forest" are trees, that is, we create a hierarchical structure) or an ability "to step back" and view the situation as one whole, that is, form a "big picture", but mainly (2) it notifies us on the inability of perceiving of such a state including indication of consequesnces. Thus, this proverb invites us not to forget to view any situation as a whole and create a hierarchical structure of the elements and find the links among the elements of this structure (including links between the levels) and the possible activities we can do with the elements. Moreover, these activities are nothing else than the creation of an abstract model of the situation, with both its static elements and their links, and the dynamic ones, the methods/activities we can perform. However, before we embark on performing the above-mentioned operations with the elements, we must

---

[40] You can substitute the word "rule" or "rules" for the "differencial equation" anywhere in this book and, similarly, you can substitute the word "environment" for the "initial and boundary condition" and then you get, what the author intended it.

recognize and assign to the elements the knowledge of the particular human, and that is an extraordinarily complex task (see chapter 2.3). We mean manipulating here the notions and symbols at a higher level. In the basic form, such a task is manageable even by animals – they also create a symbolic picture of the situation – environment, prey, danger, et cetera.

Another example is the proverb "the pitcher goes so often to the well that it is broken at last". This proverb has practical meaning when taken as a message. We all feel that pottery is not a strong and resistant material so that the message is clear: it will eventually be broken. However, the real magic of the sentence (proverb) is its generalization to other situations describing the same link between the elements. Moreover, these elements may be both material (whatever holds for pottery will also hold for other materials that may have some properties vastly different from pottery but in principle we may speak about material fatigue, micro-fissures, rheological properties, and other), but first of all it will hold for non-material elements, like human patience, for example.

We can show other contexts by using a proverb with a similar meaning – "the last straw that broke the camel's back". This proverb has in substance a similar meaning to "the pitcher goes ..".. This meaning is not exactly the same, it is only similar, we can nevertheless show here quite well that there is sort of "deeper content" in these two proverbs. I claim that the reason for this deeper content is that the basics (the deeper content) of both proverbs is just one rule, which generalizes both proverbs by putting them into a higher level of abstraction, which fully comprises both proverbs. I will try to elucidate further in this chapter how to capture two rules in one common rule.

Thinking and activity of a human being are based not only on getting concrete experiences (as an assembly of a large number of static events) but first of all on learning from these concrete experiences, that is on a generalization of these experiences so that we can handle any new situation that we have not yet lived to see. If we can find similarities with other situations we had seen, and we can find deviations from the experience, thus we will be able to solve the situation based on our experience. It is similar as if we put many points (concrete experiences) into a chart, draw a curve of a concrete equation representing the points (a rather simple one; for example, a straight line, which is a graphical representation of a linear function). Once we get this ability, we can understand and get our bearing quite easily and fast in practically an unlimited number of similar situations.

It is however still a simple description of situations which are much alike. I mean for example the relationship between the speed of a human on own power and the time for which it is possible to carry out such a moving activity (walking and running). We are, for example, capable of walking for a very long time without exhaustion (provided we have enough to eat and drink and can ingest during the walk). Were it not for the necessity of sleep, we could, in essence, walk without limit. As soon as we slowly start running (at, say, one-third of our maximum speed) to increase the speed of the movement, after some time and even with ample supply of food and drinks and no need to sleep we will have to stop because we get tired. This time depends on the individual – let us estimate it as several hours. Increasing our speed even more (say, two-thirds of our maximum speed) will sharply decrease the time we can sustain the speed. Finally, if we increase the speed of running to our maximum we can be sure we cannot keep it for more than about one minute (this is true even for the best athletes on the planet – they run on so-called "oxygen deficit" when the body cannot provide oxygen to muscles as fast as they spend it).

The example I used is the one with which everybody of us has practical experience. We do not need to know what causes such a behavior (details on metabolism), but we can form an experience (curve in a chart) which we can use without a problem for all the cases of movement on own power we encounter in life.

A similar experience in our lifetime can be formed even for other types of movement than the one on own power (on foot); for example, for other modes like riding a bike, driving a car, by going by air, spaceship, et cetera. We always find some minima, some maxima, different course of speed decrease, maximum time, and so on. All such movements may be put into a graph similar to the one we have in our head (and hence we can even put it on paper) for the movement on own power (walking and running). We get many curves (graphs) with similar shapes. Can we estimate what would describe some curves on a common basis? It is a differential equation, which forms a set of **rules** for these curves. One such a differential equation can describe an infinite number of curves (graphs) depending on the concrete conditions (**environment**) we insert the rule in; that is, we are looking for a concrete curve describing some of our transfers. This "environment" can consist of 2, 3 or 4 numbers (equation parameters), which are sufficient for describing (modelling) the given situation.

It is understandable that the precise description of even such a simple activity as the movement of an individual by various means is much more complicated than what is in the above case. However, I simplified it to explain the underlying principles and used an activity well known to any human in the world.

We think we may now perceive the way of retaining the rules using proverbs, which describe relations between elements in the proverb. The rules then can be captured using the human language or using mathematical methods (apparatus). I expect that a proverb is just another form of notation for concrete abstract situation, which can be recorded using mathematical apparatus, which is the differential and integral calculus.

The **rules** has one great magic. Do you begin to like it now? It can describe with extreme efficiency a large number of similar cases. Of course - even a straight line in a graph describes an infinite number of cases that may happen – for an infinite number of points along the X-axis in graph we get an infinite number of values on the Y axis. This, however, is a rather simple and static affair. However, a simple notation of complex phenomena may be a way used by Mother nature to process complex tasks we encounter in life. Mother nature must do it somehow like this – for example, bending and subsequent breaking of a board, bough, stick, bridge, and so on, can be described by nothing less than a differential equation. The bough needs nothing to learn nor understand and still knows very exactly how to behave under strain and breaking. Why then should a human, a product of nature, differ from these principles?

4.2. Why the very differential equations?

Why of all the things should the differential equations be a suitable tool for the brain's activity, thus for thinking? Let us try to look how a human behaves in normal situations. We take an example from normal life and is understandable by more or less all of us – fuel consumption of our means of transportation (this example is intentionally very similar to examples in previous sections). We need, quite simply, to get from a mundane point A to an even more mundane point B ☺

Let us assume that we have a car we are used to driving with, and because of this fact we know the car well. Here we concentrate ourselves on the fuel consumption for a hundred kilometers and how fast we can get through the concrete (fixed) route. We know what that consumption in the city will be and that outside of the city we will have a different (probably lower) consumption. If we go daily to work along the same route (the distance from home is fifty kilometers, thus the total distance home – work – home is 100 kilometers), it is no problem to calculate after several trips that the car consumes 5 liters of gasoline (gas) in one round trip from home to work. We will further know that the time for the round trip is two hours (one hour there, one hour back).

If we were able to measure the fuel consumption exactly on any given day we would find out that it is not quite five liters per round trip but something between 4.9 – 5.1 litres depending on conditions impacting the consumption: air temperature (seasons of the year), road surface moisture (rolling resistance), air pressure, traffic density, random times of waits at traffic lights, et cetera We assume that the exact time of the trip is noted for every day. If we measure this time exactly for the half trip we find that it varies from 55 minutes to 65 minutes – not accounting for random traffic jams or other unusual delays on the road.

We plot the data for each day in such a way that one axis (X) is for time, and the other axis (Y) shows fuel consumption; we get roughly this result:

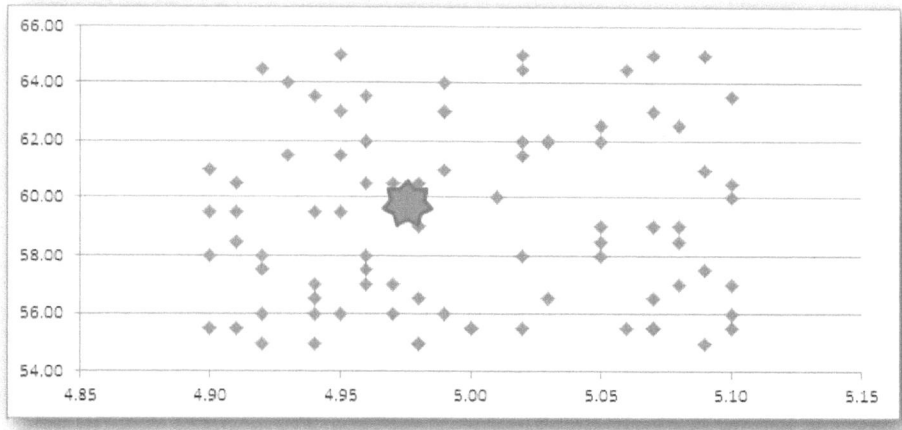

After some short time, we will know (we acquire the knowledge – either approximate or by an exact calculation) that the trip to work, in one direction, takes 59.5 minutes, and fuel consumption for both directions is 4.99 liters. This is an experience (knowledge) which we call "point experience" because as we statistically process the cloud of points in the above chart (mean value), we create the point with the coordinate X=4.99 (liters) and Y=59.5 (minutes) – denoted by a star in the chart. It is, in fact, no problem at all for humans to remember the point experience (two numbers), especially not with many repetitions over a long period.

We can continue – let us imagine we will be researching not only this route from A to B but will travel along several routes and the distance is the same – always fifty kilometers, fifty there, fifty back. We find that for different routes (all of the same length) with different landscape profiles (some are flat, some hilly, some have more traffic lights, et cetera), some trips are faster with, however, a higher fuel consumption, some are slower with less consumption. If we process each route in the same way as the one in the previous chapter, we get one point for each route (each route provides one "point experience"). We may now plot all these point experiences into a chart similar to the one above and we may get ten points for ten routes. The plot looks like this:

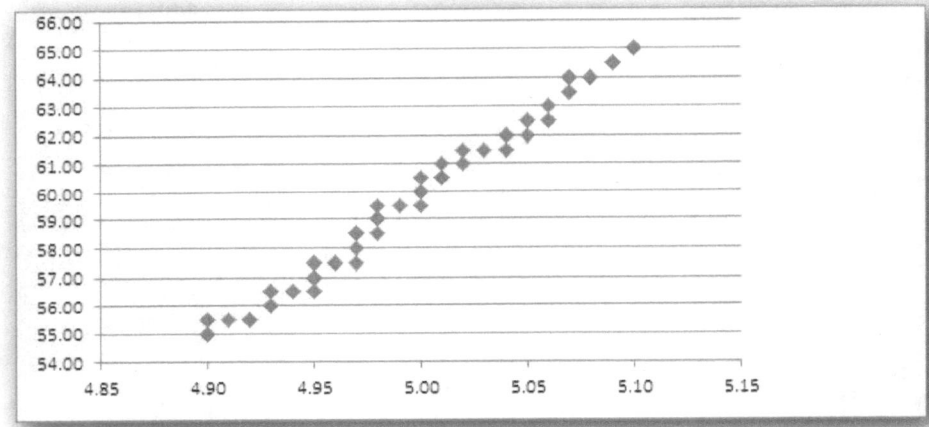

We can plot a curve through these points (a straight line, for simplicity) which summarizes possible outcomes for many hypothetical routes (we call it "curve experience"). This curve would be cut off at both ends – there would be a beginning and an end. Again, let us assume for simplicity that this curve is straight, a line segment. Such a segment is made up of an infinite number of points on it. Here we arrive at the first obstacle from the viewpoint of how the brain operates – while it is no problem to remember two numbers (5.05 liters and 58 minutes), remembering ten pairs or a hundred pairs of numbers starts to be a problem for most people (even if they would want to, they would not be able to do it). Remembering all the cases (say, 10,000 points – 250 working days in a year x 40 years, which stand for all trips to work in a whole life), or an even infinite number of such pairs where one end of the segment has the coordinates 4.9 liters and 55 minutes and the other one has 5.1 liters and 65 minutes, this is not possible for any human in the world.

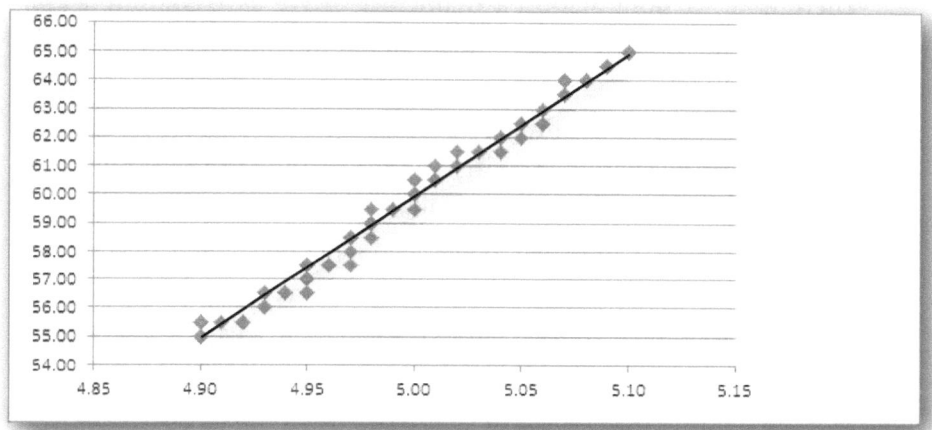

Here, however, we can be aided by a simple mechanism – a linear function, a mathematical notion that children learn in the elementary school. These children can calculate with such a function (get an exact value of Y for any value of X) and can plot it in a chart. We agree then that it is all "a piece of cake" – only we adults have mostly long forgotten it because we do not need it for life (better said: we use it subconsciously and roughly as an estimate). Such a linear function can describe many more instances than a line segment between two points with the coordinates 4.9 liters and 55 minutes, and 5.1 liters and 65 minutes. We may simply extend the segment at both ends as if it would never end – we get a straight line (from minus infinity to plus infinity – just a straight line without ends ☺).

We need to remember ONLY TWO NUMBERS to define an infinite number of points with their coordinates (every point on such a straight line) since a linear function is defined, for example, thus (here the two numbers are three and five, see below):

y = 5 . x + 3

This is the mathematical notation for a linear function which when plotted on a chart creates a straight line. This concrete line then looks like the one on the previous chart.

Where have we gone with our deliberation? We have come to a conclusion that we do not have to remember an infinite number of points, nor as few as 10,000 points, which would result from a record of each trip to our work and back for the time of our life (about 250 working days in a year x 40years of work = 10,000 trips), but we need ONLY two numbers (five and three) and a WAY (method, procedure) how to manipulate these two numbers in order to get all the possible states of the route times and the fuel consumption. This is for our brain a substantially more economical method how to form an experience for a case in life (trip to work and back characterized by the time it takes and fuel consumption) and use it for further processing to estimate some possible values of such experience.

The full mathematical formula for a linear function looks like this:

$$y = K.x + Q$$

If we plot this general linear function into a graph for say ten concrete cases of K and Q (but sufficiently different), we get ten lines:

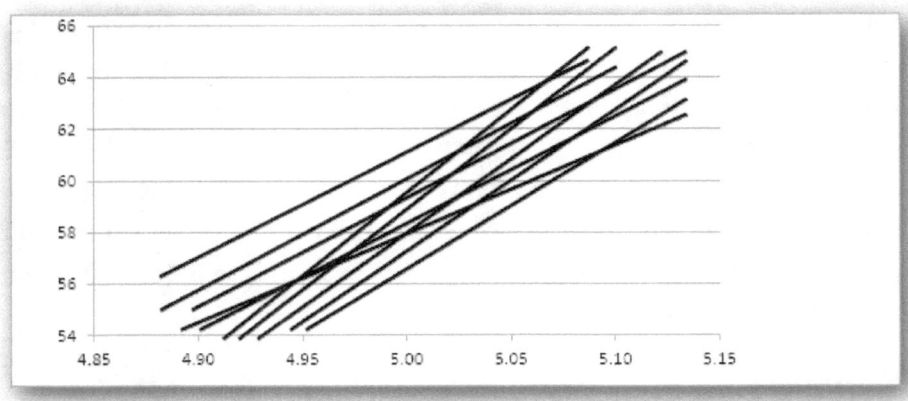

Such ten straight lines would then describe ten different routes – not only the route from point A to point B but also nine additional ones. It seems obvious that remembering ten cases is manageable for any human being, especially when they are rather similar, as in the above chart.

For all possible states of values of K and Q (all the possible routes in the world), we will get a black surface (the lines would fill the whole plane). In all these possible cases (infinitely many), one method would suffice (a linear function and its formula) but we would have to remember infinitely many value pairs K and Q – and this is again something no human nor computer can manage because of memory limitations. It is, therefore, necessary to find, for a human (as well as for a computer), another and more effective way of handling possible life situations than the one that is in fact not feasible.

Let us go on now – this has been a mental warm-up only. ☺ So far we have been talking only about lines (truly straight lines) which are an exceptional case in nature. Straight lines are the graphical representation of linear equations (as already stated - each human who passed an elementary school encountered them). Linear functions are called so because they show as straight lines on a chart. However, we encounter more often largely "curved" lines that are the results not of a linear function but much more complicated functions (quadratic, cubic, parabolic, hyperbolic, et cetera). These more complicated functions are suitable for the description of more complicated activities and situations than those given above (route and time it takes to get through it, and fuel consumption). Even a description of such a simple exercise (researching the route from point A to point B) can be distinctly more complicated – it can involve tens to hundreds of parameters at play: type of car, which engine, state of the engine, fuel, air pressure, mode of transport (road, water, air), means of transport in general, and so on. Even here it holds that we need to remember just a few values (we call them parameters of a function) and with their help we can determine (using a formula) all the states described by the function.

We can describe some different situations and states we confront in life by remembering simple formulas for functions (a small number of values). It always holds of course that by substituting concrete values (parameters) into the formula we remember, we obtain concrete results. Again, we must remember many parameters for many life situations and, also, remember what these concrete parameters mean in life. Again we are getting closer to the edge of brain capabilities as we are getting too many values that the brain should remember and finding our way among the values is, as we know, a problem for each of us.

We are coming to a situation when we have to ask: how can we handle the many equations (formulas) without loading them with the many values we have to remember? Another waypoint on our journey "point – line – many lines – ?" is the differential equation. Because this equation incorporates (infinitely) many simple formulas (functions) in such a way that for every situation (described by the so-called initial values and boundary conditions) we get a corresponding formula. A differential equation is then nothing but a **very dense** notation for many little formulas. Hurrah! This is what I, a human, need.

If we now look back again to the journey "point – line – many lines – ?", we can now supplement it thus: "point – line – many lines – differential equation", or, in fact, by "point – line – many lines – differential equation – ?". Again, we are faced with the vagueness (the question mark), of what meant by "over there"? We feel compulsory to mention the mathematical apparatus that describes the notion of "over there". I am not a mathematician, and I will only try to outline where the journey leads.

Differential equations are a mathematical apparatus which represents a large part of mathematics. There are tens and perhaps hundreds of types and versions of differential equations, each having their own special properties for solving concrete tasks. In the preceding text we described, in essence, the simple variant of differential equations – with one variable (we substitute just one value and obtain the result), but we can have such an equations of two or more variables that can describe in a suitable way a more complex task (these are called "partial differential equations"). I will use the abbreviation "DE" for "differential equations" in the following short trip into mathematics.

The notion "differential equation" is a quite a vague expression. There are differential equations of varying complexity and with various properties, but always such an equation is a generalization of reality it describes. Thus, we have an **ordinary** differential equation (it has only one unknown, a so-called variable), and a **partial** differential equation (it has more than one variable). Each of these two basic types of differential equations always has its **order**, that is a degree of abstractness — equations of first degree we can somehow understand and imagine, but a fifth-order equation is so abstract that we cannot imagine it though it may subsume the first-order equation we can imagine. We can furthermore have a **system of differential equations**, perhaps even partial DE's and all of higher orders.

This mathematical tool can describe very complex situations and relationships. At the same time, for a notation of a formula of a very complex DE we need just a few parameters, for example 10 parameters. Such a DE can describe highly complex phenomena around us and within us and uses only a minuscule part of our brain capacity. The brain can thus be highly efficient in its functioning.

I hate to do it, but I have to make it even more complicated. As is known, the world is not continuous. When we go deeper into the matter (we "zoom-in"), then after some time we get to the level of atoms, which in fact represent the discontinuous world. Let's take an hydrogen atom, which consist of one proton and one electron. The proton constitutes the atom's kernel and you may imagine it as a ball with a diameter of one meter. The electron orbits the kernel. Just to retain the size ratio of proton and electron, you can imagine electron as a ping-pong ball. Imagine the hydrogen atom as these two balls (proton and electron), which are separated by a distance of a hundred kilometers (70 miles) and there is between them only "nothingness". What is the "nothingness"? Well this is now being researched by the best world brains and so far there has been no consensus ☺.

For this reason, to research the real world we cannot use simple DE's ☺ but those that work with stepwise variables (these are called discrete). Such equations are called difference (not differential) equations. These can not only better capture the real world, but they better meet the possibilities of computer science, which processes units of information (in modern computers, one bit = 0 or 1, nothing in between). So Difference Equations better represent the real world than do other mathematical formats.

Whenever we have a concrete life situation described by a DE (without regard to the complexity of a DE) and use this DE for this concrete situation, we have in fact solved this DE with the given initial and boundary conditions (these are given by the concrete life situation). This solution of a DE is called "DE integration" using the initial and boundary conditions.

I hold that the brain (I limit myself here to the human brain) is, in fact, a machine for DE processing. In this way, we can achieve a high labor efficiency, on the one hand; and, from a certain level on, we can work with highly abstract ideas.

In neurology, cognitive sciences, psychology and elsewhere we are using the notion of a correlate. The correlate denotes two items (states, events, relations,...) which correspond mutually and are so similar as to be identical from a generalized viewpoint – these items are in correlation, they correlate. In the language of mathematics: they share the same differential equation, they are described by the same differential equation. It appears that mathematics can be applied even in areas that are poles apart from it. It is not in vain that we call mathematics the queen of all sciences. Computer science is not so much an area for an electrical engineer but rather for a mathematician. Computer science is applied mathematics, same as propositional logic or numerical calculus.

4.3. I've got an idea!

No, I haven't got an idea. This is just a way of an introduction about what to cover in this chapter. How does it happen that we get a new idea — something we haven't got before? I do not mean an idea of a genius worth a Nobel Prize. I mean such thoughts (ideas) which come up several times per day or per week (dismissing a large number of routine chores where we have no need to invent something new) when, for example, we want to simplify something.

We showed above how to describe an activity in sufficient generality and at the same time encompass a large number of cases we encounter in life.

When we pick up two (or more) activities that are described by their own rules (own differential equations) and make these two rules into a system of rules and find a solution of such a system, we then have something satisfying both rules. **This result** (new rule being more complex or more abstract than those we started with) then **corresponds to "an idea just occurred to me"**. We have quite simply a new rule obtained by combining two (or more) simpler rules. This is the way the brain can solve tasks it has never solved and be looking for new solutions and procedures. If the complexity of the initial simpler solutions is sufficient, such an idea may be elevated even to the discovery of a new (or, as the case may be, so far unknown to humans) law of nature.

4.4. What dreams did you have tonight?

Scientists have discovered that a human has, from the viewpoint of time of storing information in that memory, three basic categories of memory —

Short-term memory – here we remember an item of data or an event for tens of seconds

Medium-term memory – here we can store data or events for circa tens of minutes

Long-term memory – data or events are stored for a time ranging from tens of minutes up to an unlimited time (the whole life)

No memory medium is capable of storing what we lived through and perceived through our senses (receptors) with a 100 percent accuracy. We even may disagree what the 100% might mean. Just the fact that every eye has 100 million image sensors (receptor cells of the retina), that is it has 100 megapixels at hand (compare it with our modern cameras!) and the color depth is also extraordinary, and the eye functions with circa 25 frames per second[41]. We can add hearing, touch, temperature and other, and we are quite clear that to store data from all these senses for a whole life cannot (obviously) be done. However, even if it were to be done, the question is at which speed it can be done in order to get the data reasonably fast (the beast of prey now after me is really not going to wait until I get the necessary data to ward off the attack ☺).

We can assume from memory-management view that the brain stores data acquired during the day into the short-term memory first and then it processes them in such a way as to be able to store them more efficiently and be able to access them faster. Data from the short-term memory (immediate action – tens of seconds) move quickly and without any further processing to the medium-term memory. The time of storage here is usually in tens of minutes; however, the logic of the human cycle of day + night may suggest (I present it as a hypothesis), as does the probable computational human architecture, that this medium-term memory should be in the range 10 – 15 hours. This is the time of wakefulness (until the

---

[41] The human eye perceives 30 and more frames per second as a continuous action and it cannot recognize the individual frames (source: Techmania Science Center http://edu.techmania.cz/cs/veda-v-pozadi/798)

next sleep). In an extreme case (necessity to be awake for a longer time – several days) this time may be extended to several tens of hours (40 – 70 hours). This time is dictated by the need to sleep to clear the medium-term memory (release for next day cycle). Body fatigue also triggers the need to sleep, but it is possible that the need to sleep is called for in a larger measure by the medium-term memory being full.

This need of medium-term memory release may cause effects like microsleep when the brain tries frantically a switch to the mode of clearing the medium-term memory. It means that after some time of being full and, no new information can be put in at all (a human falls asleep or sleeps standing) or the brain tries to compress the information provisionally until it can process it fully during sleep in the medium-term memory. This provisional compression of medium-term memory places great demands on performance as (1) the brain at that time has not enough free medium-term memory that it needs for the time of wakefulness, and (2) this state causes languor, in the sense that the individual cannot function reasonably, and this is noted outside as a slow reaction.

Let us make a step – you have just fallen asleep (I do hope it is not due to this book ☺). It is a matter of fact that those body parts that rest during sleep which are linked to the physical activities, first of all, the muscles. The body regenerates to gain energy and to perform physical activity. However, probably more important and demanding activity runs at the mental level, as we stated in chapter 3.2. – Subconsciousness never sleeps. When at some time subconsciousness switches off, death ensues. Even in a coma, subconsciousness still runs, though in a limited mode. During sleep, subconsciousness then takes full control over the human (it deactivates consciousness as a function) and begins the operations necessary for proper functions during wakefulness.

There are two main reasons for processing data in the medium-term memory during the sleep – one, to release this memory for the ensuing period of wakefulness (could be more than a day ☺); second, the brain tries to mine this memory for useful information and keep substantial items that could be useful for the human in the future, and stores them into the long-term memory.

From the computational view of how the human brain functions I imagine the processing thus: the brain begins to process the medium-term memory during sleep into an abstract form in the way that brain transforms the information into experience and knowledge, that is into the form of **rules** which were applied in the given **environment** (our specific experience) However, the brain does not start from scratch ("on a greenfield") - it may do it so, though, at a very tender, if not prenatal, age. The brain tries to tie these new experiences and knowledge items to the existing ones, already recorded as rules. In this way, we can achieve a more efficient storage of information and, primarily, links are made to the experiences so that this information may be used smartly. ☺

In fact, this information processing is much more sophisticated than just merging todays three computer methods for processing of big amount of data: Data Warehouse (storage of possibly all of company data), Business Intelligence (getting a new information and knowledge from data in Data Warehouse using special methods) and Big Data (the two preceding methods amplified by several orders of magnitude — for example, continuous processing of what all users have historically exchanged on Facebook®). These areas are very progressive at present in computing; however, it is only just the beginning of the era in processing large amounts of data. The Big Data are special methods for data processing of really huge amounts of data (many times larger than we currently are using in computers around us) and anything bordering on such large amounts goes to the very edge of modern technologies.

We have in our notebook one disk, in our desktop PCs one, perhaps two, exceptionally 3 to 4 disks, but now we are talking about, for example, a million disks full of data we would like to process. These million disks need not be by far the greatest task being tackled now — we estimate that Google® has at present several million servers (more precise data have not been published) and each server has at least one disk as a minimum, but most likely it "takes care" of many more such disks (local disk array). This is directly linked with clever algorithms on which, in fact, the Google® company grew. Facebook® does not lag too far behind Google® if indeed it lags behind at all.

Let us, after the excursion into the computer technologies, return to the brain and to what happens with the medium-term memory when sleep overcomes us. We ended up by noting that the brain (in the subconsciousness) processes information from the human medium-term memory and, besides the transformation into the form of new rules in new environment, the brain combines it to the existing experience which is stored from times past in the form of current rules and its environment. From the question "what is a dream?", it is very significant to realize, what are the external manifestations of such processing?

I believe, that during the time the brain combines new and old experiences, we dream. What does such a dream mean? There are two possible modes of brain activity when we dream:

(1) The subconsciousness stores the processed rules (both new and current) in the memory which serves for the communication between consciousness and subconsciousness (they exchange here the job description and the result of the job) so the consciousness can register that there is some activity here. The activity has no meaning for consciousness as it is just a replay of newly created rules in the given environment – **we perceive a dream**.

(2) The subconsciousness verifies if it did the processing of new experience into the structure of rules accurately, that is, it must check whether after inputting the concrete values (application of rules in given environment) everything works out as it should, sufficiently fast, and that subconsciousness can move meaningful results into

consciousness so that we can use them properly after we wake up. In the time of dreaming, the brain simulates a possible situation by running it in the given environment. During that time, the consciousness can register that there is some activity in subconsciousness and the consciousness perceives that activity – **we perceive a dream**.

As the new rule is combined with logically bound older rules, it is entirely normal, that there appears in dream the experiences which we did not know that we still remember them. We cannot understand why we remembered during dreaming just the concrete story which we have percieved as to be settled a long time ago. Well, it is settled and removed from our daily activity and being, but definitely not from our stored experience and knowledge, which brain brings to mind as soon as the appropriate conditions appear. This happens because the brain verifies if it has combined the new rules with the current ones correctly.

It is obvious that the brain works at the moment of processing with our most secret wishes, too. Of course – only the brain knows them. ☺ That is the source of the meaning "dreaming of something". As our senses work on a lower level of subconsciousness (see the scheme in chapter 2.7), we can use (subconsciousness uses) everything we get through our senses – in other words: we perceive the environment. It is just a small step to somnambulism, which we can understand as not only software activity (a dream) but as the combined software (a dream in the brain) and hardware (body with its limbs and muscles).

We need here to touch upon an area all of us know well – dream books, dream interpreters, or even just our thinking why we dream such a particular dream. As I stated above, it is because our subconsciousness combined with the experiences and knowledge in long-term memory with new experiences in the medium-term memory. Subconsciousness probably moves during that activity to the memory area "readiness to pass into consciousness" the combined experience, i.e. experience which was captured long time ago. That is why we realize the experience which we "forgot".

It may have happened to you that you woke up into the "middle" of a dream, that is some external trigger caused you to wake up in a dreaming phase. When it happens to me, my concentration is down for the whole day – I feel that I am as stupid as they come. ☺ The reason is that the task that preoccupies the brain at that moment is very demanding and can be done only in the time interval when subconsciousness is in full command over the brain, even over those parts that represent consciousness (suspended at that time). If it so happens that the brain evaluates the situation as an emergency (life threatening, et cetera), then it is more important to save life than finish processing the content of the medium-term memory, and consciousness takes over by force (in other words, subconsciousness commands consciousness into an activity and tackles the issue). However, the unfinished task of processing the content of the medium-term memory, which could be completed only in a dream phase of sleeping, is such a load of both consciousness and subconsciousness that we have only a limited concentration until another sleep. Then subconsciousness "tidies up" and the concentration returns to us. For a database specialists in computer science: this is a situation similar to an interrupt when database is indexed. The process is corrupted in this instant and until re-indexed, the database searches must be done with sequential method, i.e. terribly slow method. ☺

4.5. Do you like this song and other people do not?

Music is probably a "waste product" in the abstraction capabilities of the human brain. Animals have skills to handle sounds and make them into messages for mutual communication. Whales and dogs even use ultrasound for such messages; however, sound frequency is not material for my subject matter here.

The ability to hear and process sounds is a key one for any human being from birth (this ability has the human even in pre-natal period), and we had this capability long before we were able to think consciously, let alone to think at an abstract level. This ability was though on the level of sound communication of animals.

The music we like corresponds to rules that form the basis of our thinking – for each of us. Every human has a different set of rules as this thinking basis (every human has a different experience and knowledge), and that is why each of us like different music. Even though two people may agree that they like music from the same author / singer, they will probably not agree on the compositions. It can even happen that the same composition, from the same singer but originally recorded thirty years ago and re-recorded later with a small difference, may cause someone to prefer the original one, and someone else may prefer the newer one (if he heard the new one first). Of course, it happens (sometimes on a mass scale) than many people like the same composition. This happens when the melody contains a generally shared experience translated into very similar or identical rules, which are additionally linked with pleasant moments. Or vice versa, music for a horror film is intentionally composed so that it induces anxiety to increase the film story. Music sheets and the scales major and minor exactly put down the rules which correspond to the state of mind (joy and relaxation or stress and anguish).

It is so because the melody of the respective music corresponds to the rules which are currently processed in the brain, but vice versa, too. It happens that we like the music because it corresponds to momentary mood (setting of brains environment), so this music strenghtens this mood. Similarly, but on the contrary, the music often changes our mood, as it changes the setting of brain's environment – the listening to this music forces the activation of rules, which changes our mood in the corresponding way (joyful, despondent, miserable, et cetera).

"I know you are a good executioner and enjoy your work. May I pray, however ... don't sing on the job"

Illustration © Pavel Kantorek

Music is a message to the brain at the moment of listening and the brain has to process it whether it likes it or not (you cannot switch the ears off). Every such message entering the brain is subject to an operation that translates this message into a significant form (rendering it abstract) so that the brain rules-processing apparatus may be applied. In principle, this mechanism works not only for sound (sound messages) but also for messages captured via any sense (sensor, receptor).

It is possible that the only messages that are not subject to such a procedure "at the gate" to the brain are the messages the brain itself generates – not only own thoughts and ideas but also the dreams which the brain runs at the end of sorting the new daily experience. This is because these messages are already in the form that the brain can process (composed of abstract elements amenable to brain power). The similarity with computers is again very obvious and strong.

Putting down notes according to musical scales is, in fact, describing the **environment** for specific **rules**. Individual notes of specific music scales (for example aiol a-minor scale – "a h c d e f g a" – or harmonic a-minor – "a h c d e f gis a") represent direct parameters of environment. These notes, in the form of sound during playing music, are then captured by the ear which translates them into nerve impulses so the message for the brain is created. These messages are moved into the part of brain which is responsible for the evaluation of rules in the specific environment. Different tones of different musical instruments further refine the message that music is about to pass to this part of brain. The scales MAJOR and MINOR then correspond to the mood of the individual human and the scales further uphold the mood. The environment is in line with the state of mood – the MINOR scale induces depressed and sad mood and the MAJOR scale induces joy and merriment. If the music is complemented by lyrics, then this is detached "on the way" during the evaluation and is dispatched for processing to other part of the brain (translating the text into abstract notions). However, for the complete understanding of the message, these two channels of the message merge again to complement and enhance the total.

For example, joyful music is enhanced by an optimistic text. Conversely, we may be annoyed by sad music with a joyful text same as joyful music with a sad text. Such music would confuse and upset us. We will think what such a combination means, if it is not some mistake. This need not be all nonsense – it may be an intention for another meaning that the composer tries to instill and arouse deeper thoughts by making it seemingly nonsensical by sending two messages which are contradictory. Such a processing and perception, however, already corresponds to the sophistication and perfection of the rules in the brain of each of us and the mental capabilities acquired either by education or genetically inherited. Simply put, for someone such music may be too "heavy" and simple melodies are preferred and for someone else can simple music be not satisfactory as his mental capability is used to process much more complex tasks on a higher level of abstraction.

If we are hearing a pleasant song (pleasant just to US) that we already know, its recognition may cause the brain to start processing using the rule which was alreaedy used for a similar situation. In other words: good mood from the song causes that the unpleasant reality hanging on us at the moment will be processed for a more optimistic solution (for example, we shall be looking for a compromise rather than an aggressive attack, which would be provoked by a song with an aggressive music and text). The brain simply uses the rule that is already available in this part of memory (the rule is already "activated").

There is another possibility – the brain starts to look for other possible solutions of the situation, and if there exists a non-violent solution, the brain may select it in the end as it corresponds more to the general and long-lasting preferences of the individual (the character) on the contrary to the present state of mind. The brain may though use other rules for the further course of events, but if there is time shortage the rules already in memory are finally used. The style of music may even be a sort of a filter in the selection from all the versions submitted to the consciousness of the individual by its subconsciousness. This mechanism corresponds, for example, to singing of martial songs, which excite several mental regions at once – combativeness, aggression, courage, feeling of superiority, and other, which help in the fight (it increases the militant spirit).

Similar to proverbs, music is probably an extraordinarily important element in human brain's work. After all, we all feel it, only the mechanism of exactly how it works is not clear. Perhaps it could be similar to that which I have described.

CHAPTER 5

Brain architecture of men and women

5.1. What is architecture from the computational view

What, please? Have I made a mistake in the headline? Perhaps I have in mind a comparison of architectural creations of man and woman authors? Or, perhaps, the comparison between the brains of women and men from an architectural view – which brain has a better shape?

No, I have not made a mistake. In the same way that there are designs in architecture (houses, villas, bridges, home furniture, …), both good and bad, so in the design of an information system, the significant element is the architectural design (information systems architects are one of the best-paid professions in ICT). So let us have a look how Mother nature designed human brain, and let us take this look from important social aspects – woman's and man's.

We undertake first an excursion into the domain of computers (once again ☺). Practically all computers for office and home use (not those we have in cars, airplanes or rockets, et cetera, or supercomputers that forecast weather and are a very special kind), are divided into two types: one type is the one we have for our personal use as the **personal workstation** (this is the well-known old workhorse PC – Personal Computer;

however, at this time the role is also played by a notebook or smart telephone — a smartphone), and the other type is the so-called **server** that serves several people simultaneously when they request service from it. Such servers can be found both in the enterprises, as it is a key element of any enterprise information system containing shared information (sometimes an enterprise may have up to hundreds of servers) and in homes in the form of multimedia storage or even on the web. On the web, the Google® search engine, serving hundreds of millions people, consists of roughly a million physical servers around the world forming one logical unit for searching.

"I have always thought that the brain is the most important organ in the body. Later I realized who tells me this"

Illustration © Pavel Kantorek

A FEW CRAZY IDEAS

A PC is then a computer that serves only us for our work. Often a PC requests information from another computer (server), which serves other people. Otherwise, the PC is assigned to us for our work, and no one else works on it. There is one type of PC that best shows the character of these computers – graphic workstation. This is designed not only for routine office chores but mainly for very demanding tasks such as architectural graphic designs (displaying them or computing the strength of a given structure), but also for computer games, for example.

All the above-mentioned uses of a PC have one thing in common – the whole performance is concentrated on one task: computer game, car body strength calculation, the resistance of a building against earthquakes, et cetera. These tasks put such demands on the PC performance that the user usually looks forward to a new and even more powerful graphic workstation, for an even more exact calculation of a construction design or a more realistic computer game with more details and more complex scenery. However, it always holds that the whole power of the graphic workstation is concentrated on the solution of a single task.

We have quite a different situation in servers – the server's role is to satisfy the largest possible number of clients (PC users) at the same time so that no single PC user should wait long for the request. It is usual for one server to serve, for example, 50 PC users, meaning that 50 PCs put their requests to the server. This request may be, for example, a request of a shopkeeper to the warehouse (warehouse records are on the server) if there are enough articles for sale. Another request to the server comes from the other side - the shopkeeper puts into the server information about new articles just arrived from a supplier. The server's objective is then an effort to serve as many users as possible so that each should be happy with the speed of answer.

Such a server may serve, for example, in an enterprise with many types of users – an accounting clerk creates outgoing invoices and puts in ingoing invoices, the accountant puts them into the books, the storekeeper puts in items received into the warehouse or those just despatched, the sales clerks can see how many sales items are available or expected and could be sold, and the managers see the economic results for the whole enterprise. None of these users likes to wait for answers – the server must manage all the requests so that no one would wait. The server is optimized to distribute its power among many users so all are, as far much as possible, happy.

And now the most interesting thing – with a few exceptions, both the server and the graphic workstation use the same components and even use very similar versions of the operating system (the basic operating software). The components where the server and the graphic workstation differ are, for example, graphic boards that serve in the graphic workstation to process images and graphic scenes (such a board is not at all necessary in the server because it cannot help the server successfully fulfill its role). Another specialized component is the special controller of a disk memory that is needed for the server to read an enormous amount of data from disks from which the server selects the responses for users (such a controller is not required for a graphic workstation). The main difference, therefore, between the server and the graphic workstation is then in the point that a graphic workstation (even the humble PC) tries to concentrate its power on one task while a server tries to divide its power on many requests so that all may be satisfied satisfactorily and quickly.

5.2. Men and women think differently. Why?

What is a human being from the biological and systems view? A living creature whose main role is to preserve its existence and the existence of its genus. This is done in such a way that it makes an effort to sire a new generation and ensure all the resources necessary for an offspring to continue to multiply. ☺ We have here two key roles from the systems view – provide the offspring and secure the resources. This may be done by one organism fulfilling both roles – a hermaphrodite. Hermaphrodite is not separated into a male and a female but constitutes one male-female being providing everything. Hermaphrodite produces both ovum and sperm and decides of its own will about a suitable time to procreate. Hermaphrodites are seen mostly on the lower evolutionary levels. Other, further evolved, creatures have both male and female sexes.

This division brings about the specialized roles – one role takes care **primarily** about procreation and subsequent care, the other role is **primarily** specialized to provide resources for all members of the "family" (so far we do not speak about humans solely). There is nothing revealing in this. The substantial item is that a woman (we speak now already about humans) can fulfill both her primary role (bear offspring and care for them) and the role of a man (providing resources), but a man is biologically unable to fulfill the role of a woman. The roles of men and women are rather different, and their brains have specialized architectures and functionalities for these roles. This evolutionary pressure (see also chapter 7.) has tried to adapt the human brain for the best possible fulfillment of these roles (according to the latest research, humans have been evolving for at least three million years as an independent biological branch).

Both roles (both of a man and a woman) must preserve their basic life functions that are in essence identical in both roles (apart from some differences – a woman must be ready to sustain herself and, during pregnancy, also the offspring). However, securing the basic life functions is the very basis, and the corresponding activities are very similar for both a man and a woman (not quite the same, though – obviously the priorities differ). And hence we shall not engage in these roles anymore.

The brains of men and women face two rather different tasks in fulfilling their roles in bearing children and providing for their lives. These tasks concern the best strategies for the roles.

One role (man) is focused on the resources (this is the main role of a man – after securing his own life so that he can do something else) and practically all functions of the organism are assigned to it. In the long history of humans, securing the resources was mainly by hunting, and, therefore, we will focus on this activity only (it was only a couple of tens of thousands years ago that other resources were added, like territory, its acquisition and protection, meaning a fight). Hunting was usually done in a group (it is more efficient this way) and this called for a close coordination of all hunters. Hunting, then, consisted of these two activities: (1) focusing on the pray (after hunters found it), that is its movement, intention, opportunities, et cetera; and (2) coordination with other hunters to increase the probability of a hunting success.

From the view of distinguishing the primary function of man and woman it is the most important, that a man is in such a moment not interested in anything else but hunting and to this he devotes all his mental and physical capabilities. Also, his senses are focused entirely on the information connected to this single task. Taken from this view, the whole spectrum of possible information was reduced to the narrowest range suitable for the single task (hunting). In the computer science parlance this means: all the computing power of a computer is given over to a single task and the relevant inputs and outputs. Anything that might hinder this task was suppressed (entirely or postponed). This is exactly a characteristic of a **graphic workstation** as a specialized computer.

The second role (of a woman) is focused on all the activities that connect to caring about the house and the background. In addition to caring for children, a priority for a normal woman, there was also the preparation of food, development of the dwelling, and also the relations in the group (this was and is a VERY substantial function – a social one). Importantly, none of these activities could be completely suppressed or postponed (excluding the setting of priorities). That is why the woman activity is optimized for parallel operation of tasks in such a way that the woman brain allocates its capabilities to the simultaneous solution of many tasks.

A FEW CRAZY IDEAS

It is an effort, from the view of information acquisition, to obtain the broadest possible information spectrum, quite the reverse task than with men. From the computer science view, this is an activity typical for a **server** as a specialized type of a computer – to process a maximum of requests that arrive randomly but are all needed (no client wants to wait). This is the reason women can manage what no man can: perform several activities – speak and listen or follow several independent conversations at the same time and fully participate in them, and also visually keeping track of the facial mimics and what these mean.

Scientists have discovered that the brains of men and women differ in one thing: the connection between two halves of the brain (hemispheres). This connection between the hemispheres[42] is significantly more massive in women's brain[43]. The number of synapsis between neurons is greater with women's brain, too[44]. We can view this connection of hemispheres and synapsis also from the computer science view as a bus (that is, a communication channel that exists in every computer – there are, in fact, many types of buses in computers). This confirms that cooperation of both brain hemispheres is much more intensive in women, which exactly corresponds to the parallelization of the brain's activity. Though today's massively parallel computers have a million or more processors, the interconnection among the processors is weak or none, as they work on an assignments which is subdivided into plenty of small tasks, for example the 3-dimensional space visibility task – oops, the similarity with men. ☺

The above is true for typical men and women. We know of course that (almost) no man is a 100 percent man but has (in a larger or smaller measure) a bit of woman thinking, and conversely, every woman has (in a larger or smaller measure) a bit of man thinking. It may sound strange, but it is so, and the biologists confirm it.

---

42 Corpus callosum (the source: https://en.wikipedia.org/wiki/Corpus_callosum)
43 For example see: Psychology Today (http://www.pst.wz.cz/txt/muzi_a_zeny.html)
44 For example see: http://casopis.vesmir.cz/clanek/rozdily-mezi-muzem-a-zenou-%282%29

We may perhaps know women from within our own circle whose beauty is close to the general man ideal, yet they are strict in their behavior and acting, they are uncompromising individuals with systemic thinking, who can brilliantly read maps ☺ and are good drivers, which prove their orientation in space (all these are the characteristics of a good man hunter).

Conversely, we may know men who represent the physical ideal for women (muscular sportsmen, rugged features, et cetera) but who are full of emotions, like to cook and play with children and are interested in social relations. These are certainly the extremes, but they exist and correspond to the usual statistical distribution of elements in a set — the so-called normal (also Gauss) distribution of probabilities:

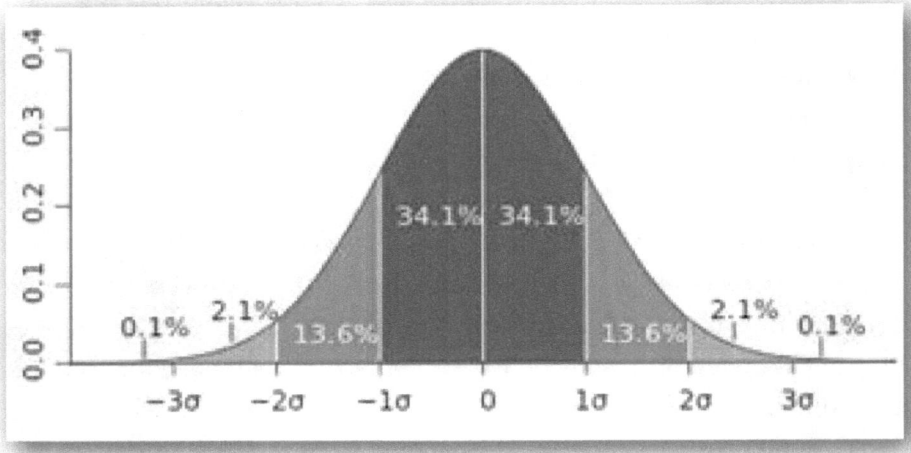

There are 34.1% + 34.1% = 68.2% of items (individuals, variables, and so on) in the range of plus or minus one standard deviation from the mean value — see the dark blue color above. If we take two standard deviations from the mean (see blue color above), we obtain 2 x 13.6% for the "tails" of the curve, for three standard deviations (see light blue color above) we have 4.2% and for more multiples of the standard deviation we have less than 0.2% of items. That is why there are so few extreme cases; nevertheless they exist.

These views on women and men may help us to behave better, and understand better, the behavior of women and men in ways they are different. None of the sexes is better or worse, they are different. They may have better or worse prerequisites for this or that pursuit, but, first of all, they are themselves. This view suggests that there is no activity cut out for men exclusively and women exclusively, but there are only greater and smaller prerequisites of any human being for a particular activity, both from the physical and the mental perception.

5.3. In comparison with a woman, a man is a simpler system

The above considerations result in one fact (a systemic, computer scientific one) which some may not like, but it is simply so: a woman as an organism is substantially more complex than a man ☺. Why do I think that this is so?

The reason is that two lives (two people), moreover strongly physically tied together, are certainly more complex than one life (one human). A woman is prepared for most of her life (after she matures from a child into a woman) and almost at any "instant" (relative to her menstrual cycle) lead "a double life", that is, to give life to another human being, at the beginning with the complete care for the new life, and she in effect doubles - she gives the child not only all substantial and material support (food, oxygen, et cetera), but first of all she launches and develops the mental life of the child (even before the birth). Can we imagine, how complex this task is?

This task is beyond comparison with the man organism, which is a mere a trigger for the creation of this enormously complex process of creating new life. Many men will not like this, but this dislike is the only thing they can do about it. In the modern society, because we are so advanced, a woman can additionally fulfill the role of a provider of resources, and this casts not a very pleasant shadow on men. However, as everywhere, it is true, that if I do several tasks in the given time, then I have less time on each task so the quality will be lower than if I have more time on each task, i.e. woman can provide the sources in high quality, but she will have less time for a child, which certainly will have negative impact on the child upbringing. I believe that this law and common sense will prevail. After all – nothing starts without "a trigger". Without regard if this trigger is known or not and what is the material form of the trigger (a test-tube or other). ☺

CHAPTER 6

You also have the sixth and seventh sense, only you do not know it ...

6.1. Everybody has five basic senses – really only five?

The well-known question "have you got common sense?"[45] will perhaps have to be re-phrased. Why? A human may have more senses than may appear at first sight. This "first sight", however, lasts not only those 100 to 200 years that humans have been dealing actively (scientifically) with the senses but perhaps thousands of years. Since that time in the past we have been feeling there must exist something like the senses and what, such a sense is, in fact. However, tens and hundreds of thousands of years ago people were not surprised that human has more than five senses because they did not worry about how many senses we had, how they were called and what was going on in their heads... It was fair enough that it worked all right. However, paradoxically, they could have a gut feeling that more senses existed, a much better gut feeling than we have today; because they knew the senses and used them.

This new look at the senses has much in common with the computer-scientific world outlook and as well as it has much in common with the human thinking and the related data acquisition from the surrounding and information sharing among other human beings.

---
45 This question sounds in Czech (author's native language): "have you got all five basic senses together?"

6.2. What, in fact, is a sense organ?

What in fact is a "sense" or "receptor organ" in the sensory reception? I do not mean here the philosophical or logical view whether this or that "makes sense", another expression for a justification. What I mean is the equipment for surrounding information acquisition. According to the classical understanding, there are five senses: sight, hearing, touch, smell and taste. There are experts who do not like the Wikipedia[46] source, but it is fully sufficient for an ordinary human being – I have adapted here the original Wikipedia definition, as medical sources give other hierarchy of sensory organs. Although the medical sources are more exact, they are not so understandable for non-physicians. According to the Wikipedia, this is the definition of a sensory body:

A sensory body (receptor organ) enables responses to the mechanical, thermal, chemical or light stimuli. We can classify them by the types of stimuli into:

- *Responding to stimuli from "**outside**" the body (exteroceptors)*
    - *Responding to a **mechanical** stimulus (mechanoreceptors) –*
        - ***Ear** – air vibration stimulates receptors*
        - ***Touch** – receptors respond to pressure on the surface and not only to the force of pressure but also to the dynamics, frequency (vibrations), and similar*
        - ***Pain** (nociceptors) – receptors responding to pain that are placed in the skin, internal organs, muscles, et cetera.*
    - ***Light**-sensitive (photoreceptors)-*
        - ***Eye***
    - *warmth-sensitive (thermoreceptors)*
        - *receptors in the **skin***
    - *Sensitive to **chemical** composition (chemoreceptors) – a cell or a group of cells able to respond to the presence of a chemical substance and its*

---

[46] See https://en.wikipedia.org/wiki/Sensory_system, text edited by author.

amount. *Nerve impulses result in the stimulation and are conducted to the higher nerve centers, for example, the brain. These receptors include:*
- **Taste**
- **Smell**
- *Respond to stimuli from "inside" of the body (interoceptors)*
  - *Sensitive to the **chemical** composition (chemoreceptors) — here the receptors respond, for example, to changes in blood oxygen concentration (glomus cells), carbon dioxide, pH, et cetera.*
  - *Sensitive to muscle **tension** (proprioceptors) — ability of the nervous system to detect changes in the muscles and inside the body by movement and muscular activity of position-movement sensation. It is necessary for the proper coordination of movement. Literally: sensation from own body, from the Latin word proprius.*
  - *Sensitive to **warmth** (thermoreceptors) inside the body — they inform that we have high temperature (except maybe for secondary manifestations like head reeling or muscle pain) and the brain would also not have information about the system (of a human being ☺) overheating, and would not be in a position to provide information to the thermoregulatory subsystem (of a human being ☺).*

The receptors always have two main structural and functional blocks: receptor cells (receptors for the given variable) and nerve fibers (to conduct the impulses).

From the viewpoint of computer science, which describes every systems processing information (computer, robot, cyborg, animal, human, space), the receptor cells are in fact specialized sensors for information acquisition and nerve cells are the connections / communication network to the place where the obtained information is processed.

This definition alone shows that there are more senses, and all could be collectively denoted as the 6th sense, or simply as another sense (feeling pain, position-sensation, warmth-feeling, ...). However, here we would like to concentrate on what is usually understood by "the 6th sense" without further elucidation. These are intuition[47] and telepathy[48]. They appear as not belonging under the category of senses, but we can expand the definition of a human sense (extension to animals and plants could be possible) thus:

"Human senses are the source of information about the outside and the inside of a human being (strange naming but it is more than just the internal bodies – see further), for the evaluation and processing so that the human can evaluate and make decisions as an independent system."

Phew! Give me pause with such a weird phrase. Yes indeed, but only after this above generalization and specification of the ambiguous word "sense" conveys a more understandable meaning to the 6th sense.

The truth is that the receptor organs are nothing but transducers that convert some physical or chemical variable into information, something able to be processed by the human "information system" that are the nerve system with the brain as the master. This information underpins human decision-making in the real world.

From the perspective of the original definition, **intuition** belongs to the internal senses as the information source is in the subconsciousness, a part of the brain (see below). **Telepathy**, then, is a special case, as the brain receives the information from the outside (from other people – their brains transmit this information about them and do this most probably by way of electromagnetic waves – see chapter 6.4.). However, reception of this information is taken care of by internal receptors (a possible way how they function is in chapter 6.4.), which transduce this information into a form to be processed by the brain. Division of senses into the external and internal ones loses, in fact,

---

[47] see Glossary
[48] see Glossary

all meaning when we generalize the definition of "sense" — it is simply now about information needed for a human being to function as an information processing system. Whether the information items come from the outside or the inside — in any case these pieces meet in the brain and are processed in the system (the human being), primarily to preserve its existence.

6.3. Sixth sense: intuition

There is probably not a single human being in the world who would not comment once in a while: **"I have a gut feeling that .."**. or **"something makes me think to .."**.. No one will suspect such a human of being a fool, or that someone supernatural whispers into his or her ears, or that he or she do not know what to do and lose the grip on himself or herself. It may also happen to us that we could not, just at the spur of the moment, remember something though we wanted it so much. However, after some time (an hour, a day or a week) the thing suddenly "comes to mind" without us even trying. What happened in the brain?

You will probably agree with me that such situations (as in the preceding section) are quite normal and should not concern us. Ok then, but where do they come from? What is this special brain activity? We are in luck because we have read those chapters of this book where I explain the importance and role of subconsciousness (chapters 2., 3. and 4.) and so we have support for our considerations.

What does it mean "come to mind" from the computational and consciousness + subconsciousness viewpoint? Our subconsciousness has learnt that some information, which we consider important (at the consciousness level), yet consciousness does not have it (i.e. we cannot to remember or "become aware of it" — only the meaning of the word "become aware" itself and the related computational brain processes are enough to deserve a scientific research and get a Nobel Prize when someone succeeds to describe these with sufficient precision ☺). This is the reason that our subconsciousness starts to look for the information because it understands this state as a request of consciousness to search the subconsciousness memory (not anywhere, but in the its own long-term memory). Such a request of consciousness to subconsciousness may also be formulated rather generally, that is partially and incompletely, and thus it takes a long time to the subconsciousness to find such an information. It must solve a really difficult task when it searches the whole long-term memory for anything corresponding to this partial (incomplete) request. This task (searching for the requested information) runs without us being aware of it (it is in our subconsciousness ☺) and as soon as the subconsciousness finds the information, it moves it to the level of consciousness (it comes to mind) and we now "remember".

This example when we spoke about one item of information (name of something or a word, synonym, et cetera), though illustrative, is simplified. The subconsciousness likely works in a more complex manner — better expressed, substantially more complex manner, as there are many such requests from the consciousness to subconsciousness running at the same time (it is a truly parallel process, not only fast switching between processes which run in series).

It is highly probable that subconsciousness monitors everything that consciousness does, but not at the operating level (what we are just doing) but at the level of possible relations and links among the information pieces that are processed at this instant in the consciousness. The subconsciousness begins to analyze these pieces; what is, however, really essential — it analyze relative to the information pieces we already have but are not aware of them (we do not consciously process them). That is to say, subconsciousness remembers even what we assume that we forgot long ago (at the conscious level — we are not aware of it), and if someone would ask us we would answer that we do not have such information, and we would not be lying — we do not have it at the level of consciousness. We can still have the information but it lies deep inside the long-term memory, and our subconsciousness must only retrieve it from the long-term memory that is low in the level of hierarchy – as if it were an archive that has a large capacity, but retrieval takes long.

In this fashion, the subconsciousness can combine recent and fresh information of which we are aware (we have it in our consciousness) with the information that we have but are not aware of it. As soon as the subconsciousness evaluates the information and finds it, it moves it to the level of consciousness where it "suddenly appears", only we do not know from where. In this way our consciousness can put a more difficult task to the subconsciousness than a mere search for a piece of information (a name), for example, to evaluate all pros and cons in a process (decision) – we say that we must "get it straight in our head". At this instant, there is a lively communication between the consciousness and unconsciousness tackling all the pros and cons as based on our experience and knowledge (coded in the long-term memory) and then suddenly, the information (conclusion) comes to our mind and "something simply tells us to do this or that". ☺

Subconsciousness represents about 90 percent of the brain computing power (see chapters 3.1. and 3.2.), i.e. the ability to process information, and it is our guardian angel helping us to tackle life situations. The animals most likely have only this subconscious ability of information processing, and we say that instincts and patterns drive them. Yes, that is to say, the animals are not aware of anything – the subconsciousness of an animal evaluates the right direction, does not think further, and the animal acts accordingly.

A human has the possibility to weigh the pluses and minuses of such information (recommendation) from the subconsciousness. A human perceives that it was not conscious, rational thinking that led to such a recommendation but that it came to the consciousness "by itself". We call such a situation intuition and that intuition tells us something. From the view of our familiar senses, it looks similar to the situation when the senses provide information for a decision (we burn ourselves on the stove). This information does not come from the material world – we do not burn ourselves by a high temperature, our nose is not irritated by a gas (gas is a matter), strong noise does not ravish our ears (the air particles are made of matter and vibrate the eardrum), et cetera, but in principle this is the information acquisition for a decision at the conscious level. In some cases (threat), this decision is made at the unconscious level – the twitch when we burn ourselves on the cooker is at the level of unconditioned reflexes. However, these are special cases when "there is nothing to ponder", and so it happens automatically. In the usual cases, we need to find out if the heating appliance is warm or cold to make a decision about switching it on.

From this view, intuition (a process in our subconsciousness) is another sense.

"Now take a close look and think. It may then remind you of more than just of a bird smashed on a window pane"

Illustration © Pavel Kantorek

6.4. Seventh sense: telepathy

Another domain so far little researched is telepathy – passing on or perhaps "just" reading thoughts over a distance directly between two persons. Some people even refuse to admit such a possibility, because they do not have this gift, so they understand it that it simply does not exist. Even the scientists often opt for such an attitude and their only support is – their authority. This attitude is of the type "when I do not know it, and I do know a lot, then it does not exist". This is a completely unscientific attitude. In this way, we can condemn several scientific fields that were originally in a similar situation. Let us be reminded of the basic principles of this book:

> **The only limitation I will respect is that a given idea should not contradict the laws of nature.**

and

> **Everything not contradicting the known laws of nature is possible.**

The moot point is that fairly often there are cases in the human society that could be explained only by direct communication among the human brains (such a case in animals can be described only with difficulties because we cannot communicate with animals to describe such a case). That is, we ask the question here – if the brains of two individuals communicate (though only in one-way – one brain reads the information from the other one), what would the mechanism be to make it possible? We shall describe here one such mechanism as a hypothesis. Sometime in the future it may be shown that the principle of telepathy is different from the one I have introduced here; this, however, changes nothing in the statement that telepathy may exist.

We understand telepathy as the ability of communication between the brains of two humans, and either in one way – one human reads the signals broadcast by another human – or in two ways – both brains read each other. It has happened to me repeatedly that I remembered a friend, a colleague or a partner, and in the next minute I received a text message from him on my cell phone. The minute is important here; were it, for example, only two seconds, I could feel the phone vibrations and quickly remember that I forgot to give him or her something. If it is a minute then in all probability she or he thought about me when he composed the text message, and their brain transmitted a signal (because he or she thought about me and only about me), my brain got the message, and I remembered. This "remembering" however was triggered by an external signal. Simply stated; the individual in question came to my mind – via telepathic communication.

Here we should, in fact, be careful if telepathy is a matter for the human brain only or whether the animals use the mechanism of telepathy as well or even more intensively. Humans started to use smoke signals long ago, now we have telephones and the Internet, and so there is no need to engage in communicating thoughts in a "dull" and antediluvian manner. I think that telepathy was much more developed in humans in the past and was practically irreplaceable and disappeared by dint of not having been used so intensively and so became stunted (how is it with our disused muscles? ☺). Therefore, the surviving telepathy capabilities in some individuals could have been considered as witchcraft in the Middle Ages, and the proponents were liquidated.

What are these communication cases we are speaking about here? There is a group of people in the world whom we can call healers. It is not important whether we call them healers, shamans, druids or something other. The point is, they all share the ability to establish a diagnosis without any instruments or tools, without penetration into the body or without any similar approach of the "classical medicine". They can often establish a diagnosis before the physicians can do it using the classical procedures. What I have in mind are the true healers where the results are scientifically verifiable (for example, a healer may diagnose a tumor by some "unscientific" method, or another so far undiagnosed ailment, and a doctor/surgeon subsequently confirms it). On the other hand, I do not have in mind those quacks whose activity produces no provable results and the whole healing is based on faith. The approach "your faith heals you" is, of course, no nonsense from the view of computer science – self-healing abilities of an organism cannot be denied, and the mind can launch these (self-) healing processes by itself because the body and the brain know best what is wrong with body (or the soul). Everybody else can only learn this in a mediated way. We do speak here about true healers who can say what is demonstrably wrong with our body or which disease grows rampant here but lies dormant there, without being able to show how they arrived at this conclusion. If somebody admittedly does not know to heal a disease, but he knows to start or unblock self-healing abilities of an organism, he is not a quack.

Another area that may appear as nonsense in the beginning is the ability to cast an evil eye. In the past I also looked at this matter with despect and condescension; however, only until it dawned on me that it is possible (from the computer science view). Casting an evil eye could be in principle "only" the ability to send a cue to my brain (subconsciousness) what to espouse as right (or, send a static piece of information which the subconsciousness espouses as its own) and thus influence through my subconsciousness my conscious behavior and the feeling that I cannot defend myself. ☺ However, let us have a look at a possible technical explanation how telepathy can function.

A technical question first: what are the frequencies of the electromagnetic waves with which we know to work at present? What I have in mind are the waves where we can either generate or at least measure the frequencies. Perhaps some of us may still remember (we had been seeing it for many long years on the scales of old manually tuned radios) that a radio using long waves (LW) works with frequencies (of electromagnetic waves) in the range from 30 kHz (a kilohertz = 1,000 oscillations per second) to 300 kHz, medium waves (MW) radios work in the range 300 kHz to 3 MHz (one megahertz = one million oscillations per second), further we have short waves (SW) 3 to 30 MHz and finally very short waves (VSW) with the frequency band 30 to 300 MHz. Let us say that the above mentioned 30 kHz are close to the audibility threshold of humans – a healthy ear can hear high tones about 20 kHz and the lowest tones are close to 20 Hz (oscillations per second). When we take a close look at the woofer (low-frequency speaker) membrane, we can even watch how it vibrates at 20 Hz.

Television transmissions (both analog and digital) operate in the frequency band 300 MHz to 3 GHz (billions of oscillations per second), same as the cell phones, WiFi, or GPS (navigation). This band is also populated by modern advanced PCs, smartphones, and tablets – their processors operate with the frequency of roughly one GHz to four GHz.

Even higher frequencies are represented by the less current technologies, but in spite of it we do encounter them fairly often – for example, satellite communication (3 GHz to 30 GHz) or radars and airport radiolocators (30 to 300 GHz). All of these technologies are quite mundane today and not only are we not interested in them; we do not even know why we should be interested. They are routine everyday matter. Waves with the frequency 300 GHz have the wavelength of one millimeter (frequency and wavelength are the normal characteristic properties of waves – they are dependent variables so one of these two would suffice yet we usually state both). For comparison – long-wave frequencies have wavelengths between one to ten kilometers (hence they transmit well through the atmosphere, and such radio stations can be received even thousands of kilometers from the transmitter).

A FEW CRAZY IDEAS

Why is a wavelength so interesting? Unlike the wave frequency alone, it can better help to explain how we are doing with the gradual increase of frequencies (and shortening the wavelength). That is, above the magic line of 300 GHz, the waves that we neither see nor feel (at least we think so) become the infrared radiation that we also do not see but can start to feel – it is the thermal radiation. Let us put in, for the sake of completeness (it will come handy in a moment), that the infrared radiation with a wavelength of one to 0.00076 millimeters (760 nanometers = nm) has a frequency of 300 GHz to 400 THz (terahertz, trillions of oscillations per second). Next after the infrared radiation we have our "good friend" – the visible light ☺. The visible light has a wavelength of 800 to 400 nm (the frequencies are 400 to 800 THz). I need to continue a little in this diversion, which may be a bit boring, but we need to get to the place, where we do not know and cannot do ☺. So please bear with me for just a while.

Next after the visible light we have the ultraviolet radiation (wavelength 400 to 10 nm, frequency 800 to 100,000 THz). Henceforth we shall be using only the THz unit as it is somewhat "human" and imaginable – in economics we encounter the amounts in thousands of billions of dollars (debts of banks or governments), unlike all the prefixes like peta-, exa-, zetta-, and yotta- ☹. We encounter the ultraviolet radiation most often when we sunbathe and need to protect ourselves against the sun radiation. It is this sun ultraviolet radiation that causes both a nice tan and skin burns.

Just next to the ultraviolet radiation, there is another "old friend" – the Roentgen radiation, sometimes called X-rays (wavelength ten to 0.1 nm, frequency range 100,000 THz to 240,000,000 THz). We know it from the hospitals when a physician examines our broken arm in detail.

And now the grand finale is coming – gamma radiation. Here the human knowledge and awareness end. This radiation has the wavelength shorter than 0.1 nm (more precisely 124 picometers) and the frequency is above 240,000,000 THz. We note that the upper limit is not stated, which means that for radiation with even higher parameters we have no more detailed divisions and names (well, perhaps the physicists, chemists or astronomers in some top laboratories have the names, but almost no one understands these). It is as if some primitive tribes of old times, whom the modern humans would meet somewhere in the virgin forests, would count till ten, and anything above was just "many" ☺, so with us, anything above 240,000,000 THz is "many". Among other things, this is also because we are in an area of radiation that is linked with processes at the level of atoms and the elementary particles making them up.

From the viewpoint of the consideration that comes up in a short while, it is essential to know, how we can work with these frequencies. We mean by working not only being able to generate such frequencies but at least to measure them. For example, a top company in the field of professional measuring devices – Keysight Technologies® (a reminder: the company was earlier called Agilent® and even earlier, before a spinoff as an independent company, it used to be a unit of the Hewlett-Packard® Company, until recently an example of the measuring equipment and medical electronics of the highest quality) – deliver commercial measuring and analytical devices for the highest frequencies in the order of tens of THz. Perhaps there, somewhere in a military scientific lab, we can find devices for higher frequencies. More we cannot even measure – we are only able to detect that it is coming from somewhere (for example, from space).

What is all this massaging with hundreds of millions of THz good for? Here we have one essential thing that relates to it quite well - we can now return to telepathy. ☺You may have commented to yourself already that I cannot hold on to a single idea ☺ and suddenly I am jumping from telepathy to a broken arm at the surgery to the astronomers. Such is not the case. I wanted only that we would move through the "good old and known" regions and the frequencies tied to them. Then all of a sudden, we enter the region of these "many many" and we are at a loss how to continue. It is fortunate that this "we do not know it" happens in a region that is not quite out of our reach (the region of the atoms). We are capable of doing some chemical reactions that are at the top of our technological knowledge and skills – we can break a water molecule into hydrogen and oxygen atoms. Or, we can produce methane (CH_4) from hydrogen and carbon dioxide (CO_2). We can do it by using a fairly large amount of electricity, high temperatures (over 1,000 degrees Celsius) and high pressures (milions of bars). However, any common plant can do the same using photosynthesis and does not need any heat nor any high pressure, let alone electricity. ☺

We can now proceed from plants to humans. What are the consequences for humans here? Nothing good. Well, in fact, yes – we are in need of great humility toward nature. And one more thing - we have the gut feeling that what a plant can do in photosynthesis, perhaps we can manage the same complex task as well, or even a more complex one. And we do manage such complex tasks. After all, the cell reproduction using DNA is, in fact, a molecular operation. And the body creates building blocks at the atomic level. No heat, no explosions, no extreme pressures. We simply see new cells growing peacefully, with all their functional differences, and this process goes on at both the molecular and atomic levels. Some of us use quite regularly food supplements – these should complement, among other, also the basic elements for our bodies – iron, fluorine, iodine, cobalt, copper, magnesium, manganese, vanadium, zinc, aluminum, boron, molybdenum, silicon, chlorine, et cetera. Our body can use materials right at the atomic level and, as I stated, no extreme temperatures and pressures are required, nothing explodes nor smokes. Our body does know how to work with materials at the atomic level.

I am aware I am repeating it over and over several times already. My reason is to initiate a change in understanding of what the body can produce (for its existence): as our body can work with materials at the molecular and atomic levels, then perhaps it should be no obstacle to manufacture a miniature and extremely sensitive detector of very high frequencies and transmit information using, for example, frequency modulation (or something else). The individual cell of our retina is able to detect just one photon, i.e. the energy on the level 10^{-19} Joule. For example, using one atom (or a few atoms) of some element or compound may be used as such a detector. And now we shall return, albeit in a roundabout way, to telepathy:

Telepathy could use such frequencies that we today can neither generate nor measure (furthermore at the extremely low power on the level of one photon processed by one atom) and thus the brain could use these frequencies for its transmitters and receivers in inter-brain communication, and with a very low energy consumption and low noise at that. At the level of waves of the individual molecules or atoms, we can operate not only using frequency modulation (a carrier wave is modulated and codes also for another signal – radio transmissions, for example) but directly at the level of the resonance frequency of some device uniques for each human (brain) same as the fingerprint is unique. Such a "smart antenna" (detector or transmitter) in the length of 33 molecules or atoms (or similar elements) is capable of differentiating 2^{33} (8 billion) of different resonance frequencies, that roughly corresponds to the human population on the planet. However, there is no obstacle in producing such a device (to receive and transmit waves) in the length of 66 (twise as much as needed for actual human population on the planet) or 128 atoms or molecules. This "large" device could possess a property quite substantial for our hypothesis – it could be so small that we have not yet discovered it, or, we have discovered some cluster of a few tens or hundreds of atoms or molecules (in the brain cell), and it did not occur to us that it could fulfill such a function. May I repeat – this is a hypothesis. Nevertheless, even the Higgs boson (a theoretically predicted elementary particle) remained just a hypothesis for decades. It was predicted in 1964 and discovered only in 2013. Ours is a bold comparison but as I wrote in the Introduction to this book: **Everything not contradicting the (today) known laws of nature is possible.** We can imagine such a device in theory. The issue is to know what to look for.

6.5. Moreover, what about the aura?

It is often stated that so far no organ or receptor has been found, which would transmit any telepathic signals (indeed, no such signals have been received). Let us now take a look at something that is similar to telepathy with about the same degree of controversy – the aura or energy envelope of a human, as it is usually described. Let us forget about parapsychology, theology, shamans, esoterics, chakras, and other – I do not disparage them; however, without further details for research they are difficult to grasp. Let us take a look at the (possible) existence of the aura from the scientific corner and, also, at the possible technical substantiation for the aura.

It is today quite common to use the methods called ECG (ElectroCardioGraphy – of the heart) and EEG (ElectroEncephaloGraphy – of the brain), which are used in any district hospital. What are these, in fact? These are the electrical signals generated by organs in our body (in this case, the heart and the brain), and we have been able to measure these signals for some time now (for decades). Measuring the electrical signals in the body (time changes of the electrical potential due to the heart or brain activities) is much easier than measuring the magnetic fields generated by various body parts or organs – heart, brain, individual nerve axons. Just for the medical uses of biomagnetism have appeared applications equivalent to ECG and EEG – MCG (MagnetoCardioGraphy) and MEG (MagnetoEncephaloGraphy).

Let us make a small overview of some magnetic fields as known by today's science (some have been known for some time, some are the results of scientific research dating back only 5 to 10 years):

Magnetic field of the earth is 50 micro teslas (μT). Its changes over time are about 10,000 times weaker than this value, and so the changes in the earth's magnetic field are in the order of 10^{-9} teslas (nanotesla, nT). Even these minor changes of the earth's magnetic field are still about 1,000 times stronger than the magnetic field of the whole brain. In other words – the brain could be able to detect changes in earth's magnetic field that are 1,000 times stronger than the magnetic field with which our brain operates or which it generates.

The weakest magnetic field generated by the brain which can be measured by modern instruments is one femtotesla (10^{-15} tesla), and it is 10,000,000,000 times weaker than the earth's magnetic field). This magnetic field is at the limit our present ability for measurement, but such a magnetic field is generated by a simultaneous activity of 50,000 neurons (brain cells), and so an estimate for the magnetic field of one neuron (a cell only, or an axon) is about 10^{-20} tesla. This value, I hasten to add, is the result of the measurement by instruments and subsequent calculation, which means that such a field exists and is not a mere theoretical assumption, as had been the Higgs boson for decades before its discovery after some thirty years.

The first instrument to measure the brain magnetic field is the superconducting magnetometer (SQUID – Superconducting QUantum Interference Device), which can measure the magnetism of about ten aT (10 attotesla = 10^{-17} tesla). The simultaneous activities of 50,000 neurons stated above represent about hundred times stronger magnetic field than this 10^{-17} tesla, and so it is quite possible to use the SQUID for the measurement. By the way – the SQUID has been known for about 30 – 40 years and if someone claims that it is possible to measure the magnetic field of a copper cable carrying about a thousand phone calls from a satellite at about 200 – 300 kilometers above the earth and that it is possible to decode all these phone call then please believe it. This is how the SQUIDs were used (are still used?). ☺ Modern science, however, has moved on.

About ten years ago, the so-called SERF magnetometers (Spin-Exchange Relaxation-Free) appeared[49]. These have the capability to measure even lower values of the magnetic field than the above given ten attoteslas (10^{-17} teslas).

Where do we want to get with this exercise? Human-made instruments do not so far work at the level of reading the magnetic field of individual neurons. The growth of every human cell, based on DNA, runs at the molecular level. An ordinary human cell must, therefore, be able to manipulate matter at the molecular level and perhaps even at the atomic level (our food provides some basic elements to the body and hence the human organism must be capable of processing them). If this is the case then there can be no issue for the human body to create a very sensitive detector of the magnetic field, with much higher the sensitivity of a SQUID or SERF (the simple calculation above shows that roughly 1,000 times more sensitive). As we shall see below (chapter 6.6.), already now, our scientists know that some birds have in their eye a special molecule (CPF – Carotene-Porphyrine–Fulleren), and they can thus probably feel the earth's magnetic field (10^{-5} teslas), including its variations (10^{-9} teslas). The CPF molecule operates under the mechanism of quantum physics, but it is rather large. Similar smaller molecules (discussions are about so-called radical pairs that work on the principle of electron spin detection, which is the direction of electron rotation) could be able to read magnetic field at the level of 10^{-10} teslas.

This brings us to the conclusion – the aura that has not been measured so far may be the whole magnetic field of a human being and at present we know that there are ways in nature how to detect the aura at this intensity. Since we know that birds can see (using the CPF molecule) a magnetic field, we can accept the theory that some people can see the aura and discern its layers, provided we can find the corresponding detector in the

---

49 See http://www.nature.com/nature/journal/v422/n6932/full/nature01484.html or http://physics.princeton.edu/romalis/magnetometer/ or https://en.wikipedia.org/wiki/SERF

human eye retina. Every cell in the human eye retina is extraordinarily complex; its activity is extraordinarily complex and so far not understood and hence there can be a hidden mechanism to detect weak magnetic fields (one suitable molecule is enough) and measure the aura. I am looking forward to the time (if it arrives) when someone can prove this — so far we have only a hypothesis. People will then wonder how simple it is, why it did not occur to us earlier, and children will be bored in the schools learning about it, in the same way as they do today learning about DNA. ☺ How many children are interested today in DNA? When DNA was discovered, and its mechanism was described, not only the lay people, but even the scientists stood in silent wonder.

One small note at the end of the aura consideration: the sensitivity of the magnetic field detectors can be so good that we can learn what happens in our immediate surroundings (up to half a meter from the body). In the case of the aura, it may be the presence of another human being, an animal, a thing, ... All this could be indicated by a distortion of the human magnetic field (the aura?). All of us know the encroachment of our personal space, do we not? Why should this encroachment be detected only by sight or by any other classical sense only?

6.6. Other senses?

One of the other senses (the eighth one ☺) could be the **empathy**. It is a purely software sense as is, in principle, *modelling* of the mental functionality of another human with the goal to anticipate the behavior of that human. Consequently, I can optimize my own behavior for the strenghtening of social relationship with the other specific human.

Empathy is usually understood as a desirable feature, which enables communication among people, but it is necessary to realize, that manipulation with other people uses empathy as the tool with wicked intentions.

The above mentioned modelling is an extremely high abstract task, which is so much demanding on resources, that it is probably not a task in the consciousness, but it is a task fully processed in the subconsciousness, which was asked by consciousness to process. The result of the task will likely be a detailed description of mental functionality and preferences of the given human, but simplified enough to enable the consciousness to work with it for this human.

6.7. Molecular and atomic fundamentals of future senses

I promised to describe the CPF (Carotene-Porphyrin-Fullerene) molecule that likely serves the birds to see the earth's magnetic field and get their bearings from it, and not only on the long flights of passage. In the case of the CPF molecule, we have a rather well-researched area, one of the subjects of quantum biology, one of the fields of study of living creatures and plants using the theory and instruments of quantum physics. The CPF molecule functions in such a way that in its basic state both the carotene and the fullerene share a pair of electrons which are essential for magnetic field detection (besides that pair of electrons the CPF molecule has otherwise many more additional electrons). Using light, the CPF molecule can get into an excited state when one of the paired carotene electrons "breaks away" and moves to the other end of the CPF molecule that thus becomes sensitive to magnetic field. The excited state may last only a short time but can be used to detect a magnetic field. The total process of transition from the basic to excited to basic states of the CPF molecule is very short.

Utilization of the molecular and atomic instruments of the CPF molecule type will likely lead to the discovery of other "senses" in humans (and animals), and so we can count on very interesting results in the future. We can, in principle, divide the senses into two categories – (1) senses mediating physical or chemical information (pressure, the presence of certain molecules, light radiation, et cetera), and (2) senses providing "only" just the existing information. The first group includes all of the basic senses and perhaps also the senses like the "CPF molecule". The other group may include information from the subconsciousness (processing of existing information) and also telepathy that receives or transmits existing information (from my or another brain). In any case, ultimately all the sensory inputs are translated into information for the brain to work on. So in a roundabout way we have come back to the same place – a human is, in fact, a computational (cybernetic) system, though enormously complex and entrancing as well.

CHAPTER 7

Darwin as Euclid

7.1. What kind of a nonsensical comparison is this?

How can I compare Darwin[50], the author of the theory On the Origin of Species, to Euclid, the Father of Geometry, which has been around for more than 2,000 years? Does this make sense? What do these men share?

Euclid[51] was a top mathematician and geometer of ancient Greece. His geometry has survived for 2,000 years and even today, an overwhelming majority of people who have some knowledge of geometry, work with his geometry. The basics of this geometry are described in his book The Elements[52] and compile several facts that are quite obvious to anyone with common sense, and we use them every day, such as that the shortest route between two points in space is a straight line. However, after those 2,000 years scientists considered this geometry as insufficient, especially for fields like the special and general relativity theories. Thus, it has become necessary to expand it (not replace it as incorrect) to further develop

---

50 **Charles Robert Darwin** (1809 - 1882) was a British natural scientist and the inventor of the theory On the Origin of Species (evolution biology). He based the theory of evolution of species on natural selection and sexual selection.

51 **Eukleidés** also **Euklides** as well as **Euklid** (he lived approx. 325 B.C. – approx 260 B.C.) was an ancient Greek mathematician and geometrician. He lived most of his life in Alexandria in Egypt

52 See https://en.wikipedia.org/wiki/Euclid's_Elements

considerations toward quantum theory, theory of gravitational field and other mathematical and physical theories.

If we consider Darwin's theory in a mathematical and computational way, it may seem that the theory is true yet represents too frail a fundament for something so robust as the multiplicity of life and its development into various species. I feel certain that many scientists will want to stone me for this statement, but there are strong arguments for the statement. One of the subjective arguments is the statement of Albert Einstein that "God does not play dice" (see footnote No. 61 in chapter 11.1). The processes inside the living organisms are very precise and cannot be based only on chance being the only determining force in the development of species. This would be such a simplification that it cannot work for long, but it can be supposed as one possible mechanism. It is, however, highly probable that there are additional mechanisms affecting the evolution of animal and plant species that could eventually prove stronger than the current universaly accepted mechanism – that of Darwin. A new subbranch of genetics has come into being fairly recently, "epigenetics". This subbranch begins to cautiously admit that there are phenomena that do not correspond to the strict concepts of Darwin's theory.

Many decades before Darwin, the scientist Jean-Baptiste Lamarck[53] created the first compact evolutionary theory. He claimed that, as an organism interacts during its lifetime with the environment (responds to it), it adapts, and the improvements created during its life are passed on to the progeny. Given the modern knowledge (genes and DNA, computer science, et cetera) it seems appropriate to take Lamarck's view on the development of species anew. It may be shown that he was, in fact, right and that even though his theory was buried by Darwin and his followers, Darwin's course was not

---

53 His full name: **Jean-Baptiste Pierre Antoine de Monet, knight de Lamarck** (1744–1829). He was French natural scientist and the author of the first coherent theory of evolution – the "lamarckism" – which came about 70 years before Darwin's theory of evolution. Recent scientific discoveries shows that it is necessary to review his theory as Lamarck could be right in his view on the mechanism of evolution of species; as well as Darwin.

in the right direction. Perhaps Jean-Baptiste Lamarck may once (soon) be rehabilitated in a big way. I think that among other issues, see below, Darwin completely ignored one of the basic principles in physics – the law of action – reaction[54], which had been known in Darwin's time already for roughly 150 years. This is such a robust basis of the physical pillars of the world that we simply cannot afford to ignore it. In conjunction with some contribution from computer science, new enormous vistas will open up, vistas until recently taken for witchcraft. ☺ But let us not get ahead of ourselves.

7.2. The Law of action and reaction in the evolution of species

I assume (I submit as the hypothesis) that the genom, as the carrier of genetic information in the DNA macromolecule, contains a part that is modified during life to pass optimized information to the next generation, so that it may survive. This is, in essence, sort of a "supplement" or also a "correction" for the original exact (in fact, the most detailed) specification how the new human (descendant) should look like. This is all done to improve the original generation so that the new generation may adapt to the changed conditions and maintain the existence of the genetic branch.

The Law of action and reaction could be implemented in humans not only on the physical level, as it is usual, but also in the control system of organism, specifically as the cyclical repeating feedback, a sort of "information processing law of action and reaction". In such a case the nervous system of the organism reacts to the action in the outside. Even the body of the organism constitutes an "outside" from the control (nervous) system of the organism's view, primarily its receptors (sensors) acquiring the information, so the organism reacts itself, in fact. A solution based on computer science is an endlessly repeating cycle where the operation parameters of the organism are changed in every cycle based on the

---

54 Physical law of Isaac Newton (1643 – 1727)

evaluation of new information acquired in the last cycle. Such a cycle is quite standard in the computer and cyber systems and it would be peculiar if it were not be utilized in such advanced "systems" as the animal and human ones. The system based on such a principle is extremely convergent to the desirable state of a system or organism. The mentioned cycle can be quite fast — many times per second — or quite slow — once per year. Specifically, one year could be the substantial cycle as it corresponds to the key cycle in nature. I cannot exclude that there exists the "organism life" cycle as such a cycle can be run with the intention of evaluating, optimisation and transfer of information to the next generation.

The presented hypothesis significantly differs from Darwin's theory, where impulses for change (evolution) of the organism are solely random. Darwin's theory rules out from the evolution of organisms one of the most important and powerful laws of nature — the Law of action and reaction on the level of control system of the organism (nervous system) realized by endless cycle of "self-improving".

Darwin's theory of organism evolution is based on the "trial — error" method, that is, we have a stimulus for change coming from somewhere — undefined where from, meaning at random, from anywhere, but it is important that from the outside. This stimulus for a change is quite non-systemic, a random change — for example a change in the genetic information by the cosmic radiation, which is a completely random stimulus. The individual's functioning (real life) only verifies whether this change "favors" the individual (that is, if the individual is better adapted for the future/changed conditions), or doesn't "favor" the individual (leads to a dead end of evolution). It is strongly reminiscent of one scientist's comparison — this scientist claimed that the probability of life's origin is about the same as if we would take the parts of a mechanical watch (cogwheels, springs, axles, case, et cetera), then would shake these in our palm up and down while expecting they would assemble again into a working watch machine.

The development of species according to Darwin could be compared to such an endeavor. We can just complement the cogwheels (three, say) with two new triplets of cogwheels (it would, of course, be quite capable of making up a functional watch using new triplet) and then would shake all these watch parts hoping for a better watch than the original one. It is possible in theory, and everybody can imagine it and in like manner the Darwin model is possible for the development of species. The issue is whether the species could have developed that way in the past and can develop so at present. At present, we can observe new species of insects, which are resistant to the recently introduced pest-extermination chemicals. The speed with which new species appear cannot be achieved by random changes, as the number of possible changes, additionally multiplied by the number of possible change activators, is higher than the number of possibly changed creatures who can develop in a natural way. Another element in this improbable process is the chance that this new correct version of a species, created randomly, will survive in the everyday struggle mode of hunter-pray.

I would like to be quite clear: I do not say that Darwin's view is not right, only that it is insufficient for the real development of life based on the fundamental law of nature – action and reaction.

I am trying to supplement Darwin's evolution of species with some elements from computer science. Darwin was not a computer scientist and hence he could not include these elements into his theory. Darwin also totally ignored the law of action and reaction that can be the prime mover in the evolution of species far greater than the random and chaotic steps (confirmed even by Darwin's supporters) initiated by the environment. If an element responds with a reaction to another element's action then these two elements exchange information ☺ and we have computer science again in place. Evolution of species with the law of action and reaction is a systematic and massive one (it runs in parallel for all creatures) and this provides more resistance than the random errors that by chance find such version of the organism to suit the external conditions for adaptation. Already such a concise description is bizarre. I do not claim that it does not work, but it is probably not the prime moving force in species evolution because it is too frail.

There are concepts in mathematics indicating quite well how a system (a function, et cetera) may behave — these concepts are the convergence, divergence, and oscillation.

Convergence means that a numerical series (series of values defined by a function, a formula, how to calculate the values) approaches gradually a certain value, the closer, the more terms we consider. For example, when we start to add values beginning with one and in each additional step we add one-half of the preceding number, and if we continue (using mathematical procedures), we arrive at an exact result — the number two. It looks thus in notation:

$$1 + 1/2 + 1/4 + 1/8 + 1/16 + 1/32 + \ldots = 2$$

It can be graphically represented as follows — the first ten added terms of the above series give the value of 1.998047, and this differs from the limit of 2, which is true for an infinite number of terms. We can see, however, how fast it converges to 2:

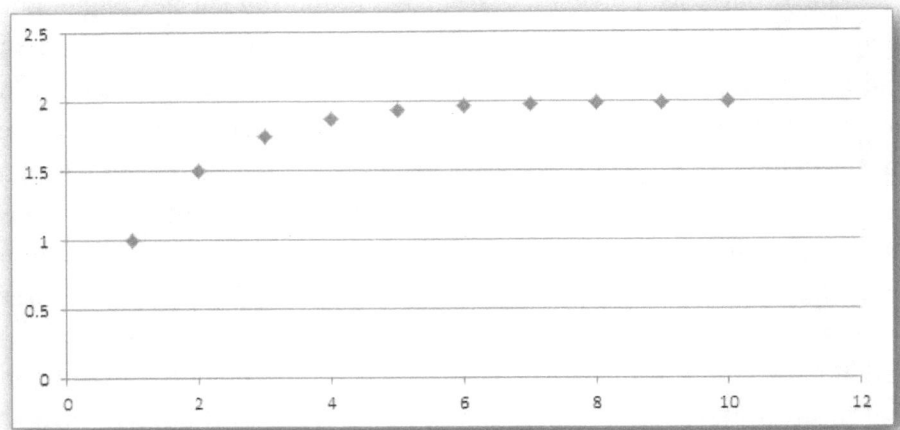

What are the cases when the value does not approach the limit value even with an increasing number of the series terms? If by adding a new term the resulting values "jump up and down," we say that the series oscillates.

The oscillation around the value "2" on a chart looks like this:

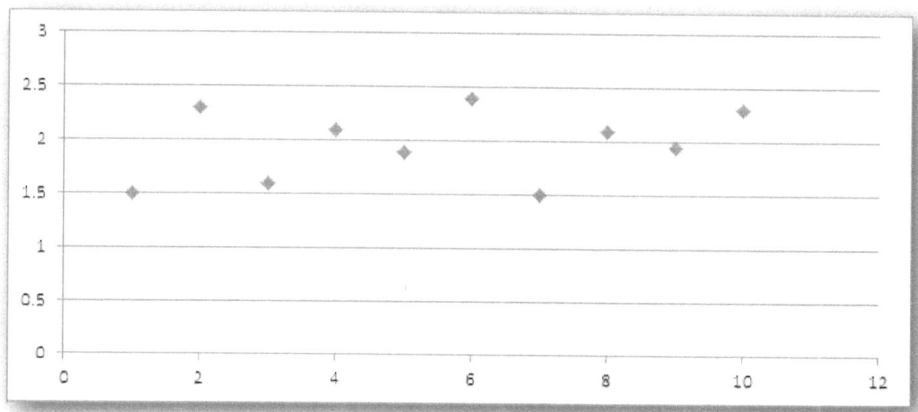

In case that the more terms we process, the more the target value differs (goes further apart) from some given value, we call it divergence, the opposite of convergence. Divergence on a chart looks, for example, thus:

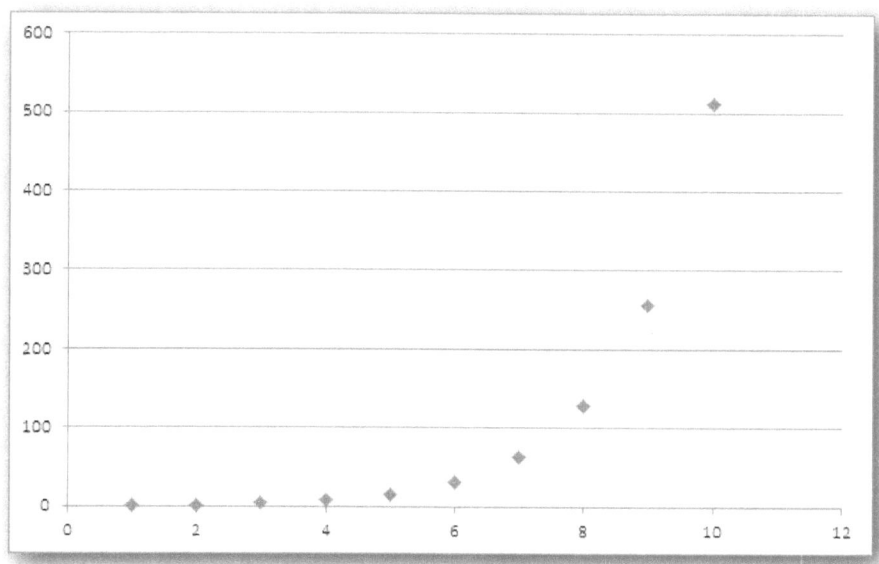

It would be good, as well as necessary, to create a model to simulate the forces specified only verbally by Darwin (and his followers), with a sufficiently large sample of elements (individuals) with a sufficiently exact description of their impacts. This would help to confirm the influence of the processes described by Darwin on the evolution of species and to compare the intensity of this influence with other possible processes. Both Darwin and his followers describe these influences as random (both in respect to time and technical aspects and thus obviously random in all aspects, that is, all are random phenomena) and even chaotic. The contingency of "trial – error" method is obviously a very weak argument for a very large system to develop and be controlled. Our system consists of roughly 10,000,000,000,000 elements (in this case, cells of the human body – ten thousand billion) and the control system consists of 100,000,000,000 (a hundred billion brain cells, neurons) and it would appear very improbable that it would sensibly evolve and **converge** toward the desired state by a deletion and addition of random non-systemic elements. Moreover, there would be, I suppose, many such trials (and errors) to make the system evolve in the proper and required direction, when the external forces may be varied (for example, the ice period, or a new predator that floated on a log from Africa to Europe via Gibraltar).

It is highly probable that such a system will either **oscillate** (in the better case, as the method is just by "trial-error") or outright **diverge** and the species will die out. It reminds me of the comparison with a scientist who searches for a cancer cure by kicking various parts of the patient's body (groins, ankle, knee, arm, head,...) and tries to see whether this triggers the process of self-healing for this particular type of cancer. The absurdity of such an undertaking is obvious. I assume that the general understanding of the absurdity of random (even chaotic) steps for the successful evolution of species according to Darwin might still take some time.

On the other hand, the attempt for a solution using the law of action and reaction brings a **strongly convergent system**. The across-the-board impact of changed conditions on many people (creatures) simultaneously generates an internal effort to change the human (creature) by the internal instruments of change; for example, by beefing up muscles or darkening skin, et cetera, for all humans (creatures) subject to a change (subject to changed conditions). Some humans (creatures) can adapt to the changed conditions faster (an essential question – why?), and they can manage the change with sufficient speed; the slower individuals will become extinct; and only the more adaptable individuals will evolve. Expecting that some random change at the level of the individual DNA units (change by cosmic radiation or nuclear background radiation, for example) would randomly target and change the right DNA nucleobases in the required way, is an act of despair rather than a usable system for the development of species: the complete human genom (genetic information) has 3.3 billion nucleobases or "genetic words" in precise sequence.

A nucleobase (the "genetic word") is made up of pairs of the DNA genetic letters – there are five such letters: **Guanine** (G) is paired with **Cytosine** (C), and **Adenine** (A) with **Thymine** (T) or **Uracil** (U). Probability that the species will evolve exclusively on the basis of a random, even chaotic, change in more than three billion elements, could be compared to an example from the watchmaking industry: a watchmaking company hires many golf players teeing off onto a green, strewn with three billion watch parts, hoping that they will somehow make a better watch by their collisions. Would you bet on it? My opinion could be guessed … ☺

Why is this chapter called "Darwin as Euclid"? I have taken the liberty to use this comparison because Euclid's geometry had remained the indisputable fundament of geometry for about 2,000 years. And then, suddenly, a time came when this geometry was not up to the standard in describing new discoveries and facts. This happened under conditions that were extreme in relation to the conditions when Euclid defined his laws of geometry.

I suppose that Darwin's theory finds itself in a similar situation as Euclid's theory after 2,000 years. I do not claim that it is untrue; in fact, I claim that it is true. Only, it is not the whole truth. The mechanism, or algorithm, in Darwin's theory is so ineffective that it would make the creatures extinct rather than evolve successfully. In a similar way, Euclid's geometry is also true now but under rather simple and constrained conditions — for example, on a plane, and at short distances and moreover without other aspects such as gravitation. No one claims that Euclid was wrong — the theory holds good even now; however, it has been necessary to develop and generalize his theory, to consider new aspects and new facts.

As I have already mentioned — I think that the mechanism of Darwin's theory works. The only proposition I assert is that the theory is not the only, perhaps even not the main, prime mover in the evolution of species. May I also doubt whether any complex system can converge to the desired solution (i.e. to create new conditions of survival) based only on a random and chaotic process.

Darwin was not a geneticist — he could not know of the double helix with the five basic letters of the genetic alphabet. Neither was he a computer scientist — he could not know the possibilities of processing genetic information in a computational system and, moreover, of processing "fixed" information (ROM — Read Only Memory) and "variable" information (RWM — Read/Write memory).

Neither was he a mathematician to know the convergent and divergent series although the fundaments of mathematical series had been elaborated already at the end of the 17th century. And so, after more than 150 years after that, in Darwin's time, they were already fairly well known. Darwin was a biologist, and his feelings guided him to find a mechanism at the level of the biological principles. It is common even at present that revolutionary discoveries are being found by an interdisciplinary team — for example, a physician in a scientific research (orthopedics) plus a physicist or a chemist may quantitatively change the view on a given problem and thus make a revolutionary discovery. Interdisciplinary approach means tackling an issue using many scientific fields (disciplines) and it is a very modern way of the scientific approach.

My apologies for the haughtiness in formulating the above-mentioned statements, but I feel that the statements underpin the direction of research on the evolution of species. Computer science is truly a devilish science – I mean, from the atheist standpoint and with due respect and humility.

7.3. The Law of action and reaction and feedback system combination

The workings of the human brain at the conscious level could easily be a computational error (nature's error or just an experiment) in comparison to all the other organisms that have been functioning for a long time (up to millions of years) without such "error". It could have happened that the brain grew due to the increase in nourishment quality (and even that could be a coincidence). The extra part of the brain may have served to store data and knowledge acquired in normal life.

It appears likely, however, that the brain has not reached the size only by the food availability, or rather an overabundance of it. The more probable alternative is that the growth of brain happened on the principle of a need to increase brain capacity and slowly and gradually (over hundreds of thousands of years) accrued more brain matter. It is like a limb (of an animal or a human) growing stronger over generations because its use proved to be evolutionarily correct for some concrete need and ultimately the limb strengthens. The growth in size and strength may be in multiples of the original size and strength. However, the limb may equally well become stunted, compare the miniature front limbs of Tyrannosaurus Rex (we cover this principle in the chapter "Darwin as Euclid"). This principle can operate like this:

The brain often began to fill to capacity its original memory of say a thousand memory cells (as an example). It may have been because, for example, a group of five hunters went on a hunt instead of just two hunters and a hunter had to remember more " friend / colleagues positions" and their combinations and this called for more memory. And so whenever there was a shortage of memory to enter the new hunting data, another part of the brain recorded a requirement for more memory, and this additional memory was added in subsequent renewals of brain tissue. For example, after some time there could be 1,100 memory cells, and this is a normal feedback mechanism. The hunter was happier with the new number of cells, and thus another feedback process occurs: this is the right step in evolution. Later, when the hunter's memory capacity reached 2,000 cells, after ten years of an active hunting life and reaching old age by the then standards, the hunter had an offspring and passed on the genetic information to grow 2,000 memory cells since it proved correct in evolution (better chance to survive and more food for the offspring). The humans could have, after hundreds of thousands of years of such information-driven evolution, a million memory cells. This number required a higher computing power (to combine and process the data), and this made another part of the brain grow whereby both the number of processors (neurons) grew as well as their combinations. That is, the computing and communication network swelled.

7.4. What does Jean-Baptiste Lamarck say here?

Almost every schoolchild knows Darwin's ideas – perhaps not in detail but he or she may know this: he is the gentleman who claims that we have developed from the apes. ☺ However, the name of Jean-Baptiste

A FEW CRAZY IDEAS

Lamarck is known only to real specialists who are involved in the evolution of species, and even these people relate the name to a tag "he was not on the right track". Why? Because Lamarck, Darwin's predecessor (Darwin at the age of twenty could have met the elderly Mr. Lamarck), pioneered a completely new mechanism of the evolution of species long before Darwin. According to Lamarck's view, an organism is subject to environmental forces in its lifetime, and it adapts to them. The improvements thus acquired in life are passed on to the offspring (simply: individuals who run for their whole life breed offspring with strong legs).

Lamarck, same as Darwin, knew neither the double helix nor was he a mathematician nor a computer scientist. His design of the mechanism exactly corresponds to the possibilities that are described in this chapter as a hypothesis.

Let us take a look at some topical facts in biology that have emerged over the past several decades. We will do it with the combined ideas of Lamarck and the genetic and the computer- scientific ones, paying rigorous attention to the law of action and reaction. There are bacteria causing diseases and against which we have invented drugs, and some of these drugs are highly effective (antibiotics). We would have thought that such a stupid bacterium is liquidated once and for all when we apply the appropriate antibiotic. And, boom. A couple of decades have elapsed, and we have a new generation (which one – the tenth, hundredth or thousandth one?) of bacteria resistant to the same antibiotic.

How come? A lucky chance (according to Darwin) intervened for the bacteria by cosmic rays, or was there another mutagen at play? (mutagen is whatever causes a mutation – a change – of the gene) Then it should have hit the DNA in the exact place and, by a lucky coincidence, in the right way to change the resistance of the bacteria exactly against this antibiotic? No one had even heard about this antibiotic some decades ago. Or, has the bacteria started to proliferate wildly and, at the same time, create random combinations of its DNA in the hope that one such a combination may cause the resistance? Please accept that this sounds extremely improbable.

We can now approach the same problem from the Lamarck-computer-scientific side and try to devise how such a defensive mechanism can work. The defensive mechanism of a bacterium will be able to respond at the moment of an antibiotic attack and before being destroyed. It will use the action and reaction law and make something like a biological imprint of this antibiotic, or even a record of the process of destruction, by the given antibiotic. In any case, during the destruction (the dying) there may be created a substance helping the surviving bacteriums to try to develop resistance against the given antibiotic. It is quite common to use such a mechanism in computer science — after the operating system crashes it creates (at the time of crashing) a record of the reason for the crash (the log). Since bacterium are known for their capacity of fast adaptation, they already have to have some mechanism of how to adapt to adverse conditions to survive.

This mechanism, to be efficient, must be fast and cannot rely on the method of trial-and-error. Quite the contrary, we can expect that the mechanism will respond actively to a concrete danger, that is, it will not chaotically make an enormous number of mutations and try all of them whether or not they are good to fight against the danger. An active response means that the bacterium, which is just now liquidated, can produce the substance mentioned above, which can help those bacterium which have survived in developing and improving an antidote against the danger (antibiotic). This is, however, already a systemic approach to evolution (not a random and passive one as with Darwin), this is a real use of action and reaction, albeit on a computational level. In other words: there is in all probability a mechanism of a defensive antidote development, triggered by a reaction to a concrete danger.

"You can't rely on heredity. I thought that my wife would cook like her mother but she gets sozzled like her father"

Illustration © Pavel Kantorek

I believe that this active approach to evolution exactly corresponds to Lamarck's theory and is sufficiently robust to ensure the survival of species, as we witness it in our natural environment.

7.5. The influence of will on the evolution of species

Our will (volition) can play a special and extraordinary role in our theory. What is will, in fact? Here we may start "laughing up our sleeve", and to consider our will to do anything, mainly to do something we do not want to do. ☺ There are several definitions of will:

Will is an aggregate of the acts of wanting, decision-making, efforts, and action.

or

Will is the cognitive process by which an individual decides on, and commits to, a particular course of action.

or

Will is the readiness to act, either conscious or enforced by circumstances.

This last definition is mine. ☺

The will to achieve something (you will define your objective) is indigenous primarily to humans, but some animal activities can be, at the non-conscious (subconscious) level, similar to the conscious human activity. In this case, we have a simplified version of the conscious human activity. The will of humans differs from the animal will by being almost always conscious and is based on a design (mental plan) of a procedure which is then implemented by consciousness. This is in contrast to the animal random search for the right way (mechanism of "trial and error"). The animal can create the very new solution of situation only as the consequence of a danger to life and concurrently none of known ways will save its life. The animal has no choice but try some new way without the guarantee of lifesaving. If the way proves to be effective (life is saved), the animal can retain it for later use. On the contrary, the human is able, thanks to abstraction, to simulate a situation of life threat in the comfort of an armchair in the living room, i.e. the human need not **experience** life threat in reality. Moreover, a human is able to simulate somebody else's life threat by "climbing into somebody's skin" (empathy), which is a high level of abstraction. A human can, using an even higher level of abstraction, look for variants of the way, i.e. he or she optimizes the solution.

The impact of will on the evolution of species can be, in principle, such that the brain should send a conscious request for some activity to be repeated sufficiently many times and always with a positive answer (satisfaction of intentions and objectives). In such a case, the biological and computational processes for organism development and reproduction, see below, can make use of this new experience and knowledge. The above-mentioned biological and computational processes mean —

- processes in the body which are responsible for the **creation of the physical constitution of the given individual** (that is, cell renewal — bulking up the muscles, for example)
- processes in the body which are responsible for the **creation of the non-physical constitution of the given individual** (for example, development and optimization of the brain)
- processes in the body which are responsible for the **creation of new cells for reproduction** (sperm and ovum — new ones will be made with the new knowledge of "this is the correct way").

I assume that the continuing implementation of this mechanism (above-mentioned three processes) will be the task of the lower levels of subconsciousness (see chart in chapter 2.6.).

Please bear with me to state two essential comments about will:

Will is a purely software phenomenon.

Our will may be the key thing for the evolution of humans as species. The will is, largely, a result of consciousness, which no animal possesses (at least not on such a high level as humans, including the ability of abstraction). And consciousness is apparently the key evolutionary advantage of humans as species. Nothing can change the fact that we will, perhaps intentionally, use our will to destroy ourselves as a species. ☺

CHAPTER 8

What were you in previous life?

"We can fix almost everything. You must tell us though what it was originally".

Illustration © Pavel Kantorek

8.1. Quackery? No, it is computer science + genetics

If there were a portion of genom in DNA macromolecule which could be changed during the life of a human being (i.e. transcribable genom part), then it would be possible to transfer the life experience of this individual to a new generation by putting it into the transcribable DNA portion. A possible mechanism could be to gradually record the new and verified life experience (knowledge and instructions like "this is the right way") as parameters of the corresponding rules[55] (in the given individual's brain). From there they are moved to the mechanism for the creation of new cells for propagation (sperm and ovum) and subsequently put into the transcribable DNA portion in the form of a rule with the environment (parameters).

It is quite possible that this feature — transcribable part — is the attribute of genom of sperm and ovum cells only, which create new life. This transcription could be extraordinarily efficient — for example, with the right design of such a mechanism, 10 or 100 parameters could be in a position to transfer the faculties for arts, mathematics, sports and other, to the offspring. If there were a sequence of a thousand transcribable nucleobases in each DNA sequence of sperm and ovum (made up of the letters G, C, A, and T — see chapter 7.2), then there would be the capacity and the mechanism to pass on the newly acquired knowledge of a human to the offspring. The sequence of a thousand nucleobases would not be even visible among the three billion nucleobases making up the full human genom. After the creation of a new life (still in the prenatal phase or even after birth), this transferred information could be gradually processed into the experience and knowledge of the new human being as we do it with new experience and knowledge during life using our senses and the medium-term memory. This would free the transcribable part of genom to make space for new data for the next new offspring.

---

55 Concrete instance of a differential equation

A FEW CRAZY IDEAS

Such a mechanism could prove to be eminently efficient for the development of species. The mechanism of the Darwin's evolution theory (natural selection, sexual selection) does not include any way of the transfer of newly gained feature to the next generation. It is probable that such a mechanism exists as the existing speed of evolution according to Darwin's theory is not sufficient to record and transfer the faculties to new generation for areas so recent that they did not exist earlier (music composition, new crafts, et cetera) and where we usually say "this is inherited", "it is in the genes" or "an apple does not fall far from the tree". All of these cases disprove Darwin and the very proverb "an apple does not fall far from the tree" is in direct conflict with Darwin's theory. Darwin's theory says that evolution prefers the individuals with features preferable for survival, but it leaves the creation of the feature to randomness (mutation based on the mutagens, which occure and apply randomly). Moreover, the new knowledges that are now indisputably transferred from parents to offspring are highly abstract, and so there must exist a far faster and more efficient transfer mechanism of acquired features than the passive mutagen mechanism of Darwin.

The idea of "previous life" is understood by a normal and mentally sound human being to be a great overstatement and nonsense. However, given our above-mentioned considerations, it may suddenly appear as a real phenomenon implemented by the instruments of computer science. Previous life would in basic outlines mean, that we have in our mental faculties (our rules in the brain) the knowledge and capabilities of our progenitors, and when we activate the capabilities, we can create the impression of having lived those lives; we bring them up in the same way as if we lived them albeit a long time ago. In reality, these capabilities were experienced by a progenitor of ours, but we live them out now as something that appeared from "some place" in our brain though not created by us – Déjà vu. Such a feeling can be formed inside us when we use this rule or the knowledge for the first time. It is a very simple and elegant solution.

If this hypothesis of such a solution would prove true, then we can give all due credits to nature. It attests to the quality of the design of the human mental (i.e. computational) activity.

A talent is just another view of the same matter — either we have the preconditions to develop further by our efforts, or we do not have them. This would prove both that we have such a faculty and also how we acquired it. In fact, the issue here is this: the preceding 20 (200 or 2,000) generations of an individual were engaged in some pursuits that biased all of them for a concrete faculty (a mental or physical one). Then, it is highly probable that an individual in the most recent generation shares these faculties. They are stored in the DNA as guidelines / program for the given faculty and the individual may, or may not, develop them further (compare the proverb "an apple does not fall far from the tree"). It is so difficult to embark on a journey in a direction for which we have no preconditions as we have to develop them from ground zero. On the contrary, when we touch upon an area for which we have the gifts in our genes (instructions/program) then these faculties fall into place very fast, and we usually call this "he or she has it built in the brain cells". This is just for want of a better name but, in fact, it is no more than a "dormant program" that is activated.

8.2. Do older parents have cleverer children?

What has been stated above results in one conclusion that may have occurred to the readers as well because it is a logical conclusion from the above-given hypothesis. If the parents do create better and better sperm and ovum already during their lives (an ovum vanishes after each menstrual cycle and a new one must come into being; sperm are produced more or less continuously based on the man sexual activity), then this mechanism would mean that an offspring born later (to older parents) would have additional experience acquired by the parents from the time of birth of the first or previous offspring. It seems to be a piece of nonsense, but, in fact, it is just a logical consequence of such a mechanism.

We can now start on the predicate that there are a thousand nucleobases to pass on the experience to the progeny (there can easily be a hundred or 10,000 of them, the actual number is not of a substance). These nucleobases are continuously stored in a buffer and then copied into the newly created cells of sperm and ovum. The brain subsequently (during sleep when the medium-term memory is processed, among other) keeps creating better and better sets of these thousand nucleobases, and these are simply copied into the transcribable portion of the DNA of sperm and ovum. Elementary, my dear Watson[56].

---

56 The statement of detective Sherlock Holmes to his colleague and friend Dr. Watson after successful solution of a tough problem ☺

CHAPTER 9

I want to be smaller, smaller and smaller, to become the biggest in the world

9.1. A muddled heading? No

Before I conjure up one possible principle of how to understand space (naturally a crazy one ☺), I need to prepare the ground for it. There are things around us which, though rather simple, have a touch of strangeness about them. One such thing is the so-called Möbius strip[57] — a thing to be created in less than a minute (we only need paper, scissors, and glue). Once we have created a Möbius strip, we introduce the Klein bottle. Will it be full of divine drink? No, but it will be full of something else (so far a secret). Creating a Klein bottle is more demanding but we can certainly imagine it.

9.2. Möbius strip

Take a sheet of paper (clean, if possible) and using scissors, cut off a strip (ribbon) about 6 by 0.5 inch. We hold it in our left hand at one end and in the right hand at the other end so that we can see only one side of its surface. Now we lengthwise twist one end by 180 degrees (half a twist).

---

[57] It was discovered independently by both the German mathematician and astronomer **August Ferdinand Möbius** (1790 – 1868) and the German mathematician **Johann Benedict Listing** (1808 – 1882).

Now move both ends closer, join them (both ends overlap) and glue them. Now let it dry and we can do some tricks with that miracle:

Möbius strip

A moment, please — where is the miracle? Yes, what you can see in front of you does have some miraculous features. Do you not believe it? Let us go ahead.

Let us first just put our finger on any place of this joined strip (of Möbius, of course ☺). We can now move our finger along the strip (in any direction) to try to go to the end of the strip. Then return to this text again, please …

So how was it? How long have you wandered along the strip? We may become aware after a short time that we cannot reach the end, simply because there is no end (nor a beginning) on the Möbius strip. No matter in which direction we start on the trip and always keep moving forward, we would not even recognize the end and the beginning (were it not for the feel of the glued ends). We need to realize of course that we can move in one dimension only on the Möbius strip — there and back. No left turns, no right turns (in practice we could move crosswise on a strip of one centimeter but this is not of interest). An ant running forward along the strip would never reach the end while keeping running forward.

I can already hear the objection: why to do it in such complicated way if it is possible to do that simply so that I glue the strip without twisting the ends so the ant can run forward on the strip and never reach the end. This is not the same. In the case of the Möbius strip, the ant runs on the whole surface of the strip. The main beauty of Möbius strip is that the ant runs at a certain moment at the starting point but on the "opposite site" without becoming aware of it. In this way, a regular paper strip (Möbius strip) can simulate one Universe where the objects lay close to each other without knowing it, while the ordinary "untwisted" strip represents two parallel world not aware of each other.

We will experience something similar when we move along the edge of the Möbius strip – there is only one edge! Now we may perhaps note why I called this tiny thing a miracle. The Möbius strip was discovered by the mathematician Möbius in 1858 (it was discovered at the same time by another mathematician, Listing, but Möbius was luckier from the historical perspective, and the strip bears his name ☺).

It is exactly this uniqueness of an only one edge that has a connection to another magic thing – the Klein bottle.

9.3. Klein bottle

As already said – a Mobius strip has only one edge. Now imagine a circle cut out from a piece of paper – how many edges are here? Only one edge (border), too. We can cut out a circle with exactly the circumference of the length of the Mobius strip edge (one edge only there) – this is a simple calculation, and I am not going to delay the reader with it. Now we can try something that should be logically possible: we connect the circle circumference to the edge of the Mobius strip (we sort of stitch them together), and as these are of the same length, it should be successful. It could be somewhat laborious, but if the circle is made from a flexible material, we will eventually succeed. If we are not successful, we can at least imagine it. What comes out of it?

We have an analog of a Mobius strip – whereas a Mobius strip is for a movement in one dimension (back and forth – as on a thread), a Klein[58] bottle is the analog for a surface. We can move on the surface of a Klein bottle, and we cannot leave the surface. We can, of course, move on the surface of a sphere without leaving it, however, we cannot get inside the sphere; with the Klein bottle we can also move to "the other side" of the surface without realizing that there is an antipode close by – separated only by the thickness of the material making up the Klein bottle. The very idea is very surprising: we are only, say, one millimeter away from something in space, but reaching it by moving on the surface of the Klein bottle may be one meter.

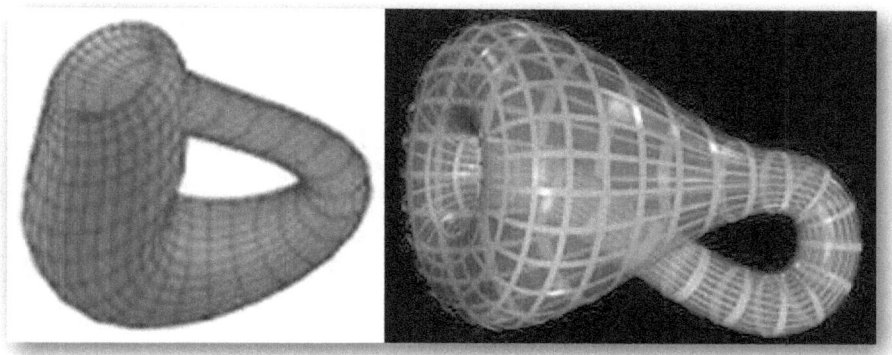

Klein bottle – different projections (source/authors: Wikipedia, VTM/E15)

---

58 **Felix Klein** – German mathematician (1849 – 1925), he was engaged in non-Euclidian geometry and theory of groups.

9.4. Do you like to increase or, better, decrease your size?

Now the main topic of this chapter:

We can move back and forth on the Möbius strip without any limits (without having a clue about the beginning and the end, and so we move round and round). In like manner, we can move on the surface of the Klein bottle in all directions (only on the curved surface) without limits (without having a clue about the beginning and the end so we can move in circles) and thus we can grow or dwindle our size (in our imagination) in space without any limits and so can grow or dwindle around and around.

What implications will this have? Imagine first that we will grow our size. First, imagine that the earth is the size of a basketball (anyone can imagine this — the downsizing is only by a factor of about ten million ☺). Let us move on — at the next "stop" in the growth, the Sun is the size of the basketball and circling around it are smaller balls (planets) with their moons, some dust particles (planetoids, comets, et cetera) and all these held together in place by the force of gravity. All this happens on a scale that we can imagine with the elementary school knowledge, even though the growth in size is now by a factor of about ten trillion (10,000,000,000,000). However, we are still only at the beginning of the growth.

We can make this route more attractive as not to become bored — who would like to wait for a year before the blue third small ball (the Earth) makes a full circle around the yellow ball (the Sun) — so we make it snappier. We speed up the orbiting times. For example, the third small ball will now circle the yellow ball in a few seconds (in our new time scale). It is still boring to wait until the farthest small balls, (Uranus, Neptune), make it around the yellow ball. Because we have just now managed it so well, we can again start to grow "a bit" and accelerate time (to have our Solar System with the diameter of about one meter and Neptune would orbit the Sun in a few seconds — again a change in the perception of time). By the way, if we take a look at this scaled-down Solar System and view it as the model of an atom (the Sun is the nucleus with the protons and neutrons, and the electrons are the planets), we will have imagined the same for what Ernest Rutheford[59], a physicist from New Zealand, got the Nobel prize for chemistry (and the discovery of radioactivity) in 1908, and thus he founded the field of nuclear physics. ☺

However, ahead of us is another journey of growth and acceleration. It will be a long one this time – unless we shorten it by letting everything grow in our mind much faster than until now. We will rather not even indicate how many times we grow – an idea is enough as the multiples will cease to have a meaning very soon – let us be surprised. We will get to the level of our Galaxy (sometimes called the Milky Way), now about the size of our stretched arms. Our Solar System is now so small that we cannot see it with a naked eye – on this scale, its diameter would be about a hundredth of a millimeter, and we do have a problem to see such a speck of dust.

"So here are the results of your psychological tests"

Illustration © Pavel Kantorek

---

59 **Ernest Rutherford, 1st Baron Rutherford of Nelson** (August 30th, 1871, Nelson, New Zealand – October 19th, 1937, Cambridge, United Kingdom) was a New Zealand born British physicist. He is considered to be the founder of nuclear physics. He researched the radioactive decay of chemical elements and presented the concept of the half-period and proposed the naming of the emitted radiation as α, β a γ *(Alpha-, Beta- and Gamma-ray)*.

If we are now in the grips of feeling unique of how big a Galaxy we inhabit, this feeling will soon pass away – our Galaxy is a small part of the so-called Local Group of galaxies made up of 42 dwarf galaxies (who invented this name? ☺) and three large galaxies (one of them is our Galaxy – wow, finally we are somewhat important). Only wait a wee bit— we are growing again and accelerating at the same time so as not to wait for an eternity before our beautiful spiral Galaxy makes a full circle.

It may be already slightly boring, but I have to keep on growing and accelerating everything around me. Let us take in one hasty sweep all the usual supergalaxies ☺ including even their massive black holes in their centers (a black hole is something we cannot see – its gravity does not allow even light to escape) and suddenly we are getting to the frontier of the known universe. We may want to get at least an idea how big the Universe is according to the present knowledge. So here we go: the light from the Moon reaches the earth in about one second, from the Sun in about eight minutes and the (currently known) border of the Universe in about 14 billion years. The earth's age is slightly more than 4 billion years, and we would have to wait a while (10 billion years) before this light from the edge of space finally reaches us. Can we agree not to wait and to go on? ☺

But now we have an issue – where to expand? In my view, there are two ways to proceed:

One possibility is that we will see no border at the distance we have so nicely described. Same as on the Möbius strip we will keep going forward without noticing that we are again at the beginning. We have reached this beginning "from behind" and now we are starting on a new round. So if we gradually expand in space, we can get to a state of returning "from below" to the elementary particles, the atoms (yes, they do look like the Solar System) and the molecules and all the way to the normal size of ours. This is how we can explain the "infinity" of the Universe: the trip along the Möbius strip is also infinite (round and round in circles).

Another possibility, which could be only a diversion on the above journey, is this idea about the edge of the Universe:

Have you ever heard about the so-called Big Bang? It was the beginning of our Universe when all matter and energy were packed into one point (no worry – even the physicists themselves have troubles to imagine it – if indeed they can at all). Then there was a bang (no one knows why), and this is the Big Bang. The whole history of space is, in fact, only an explosion with a subsequent expansion of space: space keeps on expanding, there are bangs everywhere and some nice things like the Sun and the Earth and … life originate (see later in this book).

Now imagine that somewhere there "beyond the borders of the Universe" there sits a boy (a really big one) who is playing some chemical game (or only with some batteries) and the fabulous Big Bang of ours is just his experiment that went wrong. Something went bang in the test tube (or the batteries were shorted), and the result is our glorious world. ☺ The big boy sees this explosion or the spark as a fast process. However, if we (for a change) slow things down and enlarge them, we can get into the chemical reaction or the spark, and we are back in our Universe.

Why do I call this idea a complement to the original idea of "expansion in rounds"? This is because even the boy (the big boy) can grow up continuously so he starts to return from the elementary world. This could create a hierarchical model of the infinity of the Universe. A crazy idea, to be sure.

These ideas make it possible to tame the infinity of space and elucidate what is there beyond the end of space if indeed such an end exists, and whether or not we are just running around as on the Möbius strip.

My deepest compliments, Mr. Möbius.

CHAPTER 10

Eternal question – what is love?

10.1. Nothing supernatural (my apologies)

Well, this is really on thin ice – I may be called names for technocracy, insensitiveness, absence of emotions (being an emotional idiot) and so on. But somehow it must function because it does function, doesn't it? We are a part of nature, so there is nothing supernatural.

10.2. We are an animal – only a more complex one

Whether we take the view of Darwin or Lamarck (see chapter 7.), in one aspect they are agreed – we have evolved from the lower creatures. We should not forget the basic reason of human existence, too, given us "by nature": humans are living creatures, and their main task, it seems to be, is to preserve the existence, both their existence and also the chain of existences through their progeny. The efforts of a human to find a mating partner is the third most important one; that is number three in the ranking. ☺ As we have already explained, it makes sense to find a mating partner only after my life is preserved (there is no more a danger of losing own life) and when the conditions for a long-term existence of our person are maintained (air, drinks, food). Just "to have the progeny" means not only to transfer the genetic material to the partner (a woman) to create the new life but (see chapter. 5.2.) to assure the relevant resources for all three persons, too. ☺

Humans are social creatures. However, one type of a social bond is superior to all the rest – the social bond to a partner. This bond serves for one of the most powerful human instincts (as well as the instinct of all creatures) – the instinct to propagate. There exists at least two phases of relationship – to be in love with somebody (infatuation) and to love somebody. There is, of course, love for our parents and other people (other than sexual partners, that is mating partners) but this is another kind of love. We are talking here about being in love with somebody (infatuation) and loving somebody of the opposite sex (I do not touch here upon the homosexual relations but only upon the heterosexual ones, which can guarantee an offspring and human survival in the long term).

Most of us will know this "bolt from the blue" when we meet an individual of the opposite sex, and we are so enthralled as to spend the rest of our life with him or her. Of course, it is a wonderful moment, and we do not think about anything else ☺. I believe that most of us have experienced such a moment but what is this process from the computational point? This process has two aspects:

On the one hand, this is a general aspect, that is we have inside us an encoded instinct so that we and our genes may survive. Any given individual searches for a partner to sire future generations. The partner must be such as to increase the probability that the offspring will also sire new offspring. The partner must be somehow attractive – beautiful or handsome, agreeable, able to survive, et cetera. These attributes must be taken into account because our offspring will strive for a partner in the future and hence this aspect must be universally accepted. We may, as individuals, like bow-legged girls with hoarse voices, but it is not a universally accepted norm of interest for men. Our offspring with these qualities will probably have an issue in finding a new partner to sire the next generation. Also, our genealogical branch of "admirers of bowed legs and hoarse voice" will become extinct unless, of course, our fancy in these "non-traditional qualities" will contribute to a better chance of survival under the given conditions (new ice period and similar). Then, on the contrary, these qualities may prove favorable for the survival of our generation, and as soon as this is proven, our attractiveness as a partner will rise sharply. We had, that is to say, an evolutionary advantage given by our intelligence – we were able to foresee future development. In normal life, people often make the mistake in the abilities – rather than providing resources for the family, they prefer the ability "be able to steal sufficiently enough resources without (until now) being caught and punished". ☺ But the end justifies the means so this can be the real way of evolution.

There are additional personal aspects causing that certain qualities appeal to us only – first among these are our acquired ideas about our partner. We have acquired from our parents (for a boy, a beautiful girl is the one who resembles his mother, and most boys do not realize this). Next come the features (as the acquired ideas) which appear attractive to us only. These acquired ideas are combined with the inherited ideas to correspond to the current state of society.

Love (even unrequited ☹) arises at the instant when the given individual's "partner selection subsystem" finds a person who in a large measure fits his or her ideas about a partner. At this instant, he or she "has a crush on her or him". Because we may feel that we have arrived at the end of the journey, mechanisms are triggered to lower the level of judiciousness to successfully win this partner "for a joint lifelong journey" (primarily for sexual propagation). The lowered levels of judiciousness let us into the stage "rose pink-colored glasses" (we do not know anyone who is a better fit to our idea of a perfect partner).

Another level of euphoria and enhancement of the "rose-colored glasses" effect comes when the "selected partner" requites our crush (that is, there is no period of frustration). At this instant, the mechanism to lower our sound judgment increases even more substantially (due to hormonal processes – the hope to continue the genealogical branch is one of the strongest and most elaborate instincts). In fact, a human starts to act like a cross between a drugged junkie (a piece of cake for endorphin-like hormones) and a mental case, since the usual checking mechanisms are switched off. This is normal because success in sexual propagation is the key objective of our existence as such. Such an affected individual begins to be completely uncritical toward the partner-to-be because conscience says (commands?) that the key life objective (i.e. its life propagation) is close and everything else is of less importance. An inseparable part of such behaviour and action is lowering or even elimination of vigilance, cautiousness and self-defence towards the beloved person, which explains the well-known truth that "the one, who loves more, loses more in the love affair".

There is no change here even if these two individuals have no offspring in the end (for any reason, including infertility). It does not change the perception of the other person – understandably, unless the reproductory instinct commands the enamored individual to find another perfect partner who can (or wants to) bear children. We may attach a touch of supernatural virtue to this state of affairs (infatuation). However, in reality it aims at reaching a higher goal – bear offspring who later succeed in having their offspring. May I apologize in case you may perceive this as technocracy and lack of feeling.

There is an essential, in the time interval between my crush on the given person and her or his possible (but uncertain) infatuation with me, a sharp increase in the efforts to captivate this person (me to appear the best and the most interesting) so that love could be requited (people not in love call this "behave like an idiot"). We try to look in the best possible light or a better light than we are (this is a part of the genetic strategy – we do not care so much about the way in which we win the object of our infatuation). Metaphorically specking, a farmer reveals the longing to be a seafarer[60] ☺ which may not be brought to fruition but may be significant in getting the desired person to beget the offspring. At the conscious level, we are only capable of correcting this intention by satisfying our libido (to have sex with no offspring).

Psychology teaches us that the factors given below are taken into account when we select a partner (the procedure is "from the lower instincts up to consciousness"):

- **Body** - selects by instinct (!!!): smelling or kissing (kissing is the first test of genetic compatibility, a woman verifies if she can have healthy offspring with this man). Smelling is a similar test, among other factors because we cannot reasonably "hoodwink" the smell since it connects directly, without additional processing, to the brain center – hence it is substantial whether or not the given individual has a pleasant smell, natural and not artificial (musk).
- **Enjoyment** (emotions) – selects using hormones
- **Reason** – common interests in some activities, that is social compatibility
- **Relational correction** – social status. For example, providing for the partner so as to ensure an easy upbringing of the offspring. The well-known adage, "money is the best aphrodisiac" ☺, comes into play here.

Relationship (to another human being, a personified thing, or an abstract item) is one of the key needs of a human being. We can say, in a simplified manner, that it is natural for us (regardless of our gender) to look for a friend or partner. A friend covers a certain part of the relationship needs of any human (man or woman), primarily the social part (need to communicate – asexually here, we can have a nice chat even with a person

---

[60] Folk singer Pavel Dobes in the "Something about love" song, see http://www.karaoketexty.cz/texty-pisni/dobes-pavel/neco-o-lasce-122071

of the opposite sex although we feel no desire for a sexual partnership – see below).

Another matter (higher level of human relational needs) is searching for a partner. We are inspired toward this partner search by one of the basic instincts – the sexual instinct (I consider here only heterosexual individuals as they form the prerequisite and fundamental element for the survival of the humans as species – I do not have an issue with the acceptance of homosexuality, but two men or two women can simply have no offspring of their own making.

Illustration © Pavel Kantorek

Let us be reminded of how strong these basic instincts are. We include here mainly these:

- **Self-preservation** – probably the strongest instinct since an individual must preserve its existence (must prevent death) – without

an existence it cannot strive for long-term survival (when I drown, it makes no sense to provide food or contemplate offspring).
- **Sustenance** (also "provisions") — as soon as we are maintaining existence (see the preceding instinct), it makes sense to provide the means of living (for myself and close relatives) for the energy and nutrients needed for further development. Unless these are provided, it makes no sense to consider the offspring. The same is true if we change the priority only for this instinct, pass on the genes and die. This process, however, can be a specific strategy for the survival of a tribe (we pass on the genes, and someone else takes care about the resources for the new generation's development, hopefully).
- **Reproduction (sex)** — as soon as both our basic existence (life) and the continuous existence (sustenance) are guaranteed, it makes sense to have offspring. Under normal circumstances, we take care of them by providing resources for the development of a new generation.
- **Sociability (grouping), aggression**, et cetera — these build the superstructure above the basic, (de facto animal instincts). Without the basic instincts, no organism in the world stands a chance of existence and survival.

We can imagine the search and finding of a partner without love, purely at the level of instincts and hormones (we want an offspring). However, most of us also strive for the best relationship with the partner. We are looking at this partner as so much of a perfect one that a feeling may arise "I do not want to live without this individual, I am inexorably attracted to her or him", and this is infatuation, which will (maybe) segue to love.

Now we can return to the question in the title of this chapter: **what is love**. I will try to answer this question, as before, from the computational view. After all, a human is nothing more but (excluding the organized matter) a computational (cybernetic) system and the soul is the software and data.

In our mind (psyche), we have the criteria to evaluate relationships. In this view, it can be even a negative relationship (hatred). We are "taking a view" (an opinion) of every human we encounter. Very important is the first impression that has a key and long-term influence on our relation to this human. First, we create a computational model of this human as a character (who he or she is, what he or she does, says, ...) and practically at the same time (the first impression at the physical level can be made in under one second, the first impression about the character takes several tens of seconds) we can create an opinion. This opinion is entered into the **relationship table** of this individual, and the table contains all the important features of this particular relation. This table is assigned, once and for all, to this individual and we will be only adding to it and make it more precise (change it). In fact, based on the first impression, we can assign additional functions to the individual that are essential for the relationship – we can, for example, say, "I need to watch out for this guy" and this is nothing else but a software function. This function will more strictly evaluate all that the individual does and can relate to us (even though only in the future – I simply need to watch out).

This relational table contains both the features that are communicated from our subconsciousness and features of which we are made aware by our consciousness (reason). We can simplify the ideas about this table by imagining it as a stick with the individual features written on it (their number may vary). One part of this stick is immersed in the subconsciousness (water) and the other part is in the consciousness (air). We can then have such a "stick" for every individual we have encountered, however shortly. Sticks of individuals we have not met for a long time are stored in the long-term memory.

This model of the external evaluation is common to all animals that, however, judge the environment in a much simpler way (non-socially).

Their primary aspect is — a mating partner — friend — prey — enemy, basically in descending order of positiveness in the qualities of relationship. This level in the animals corresponds to the part of the stick (see above) that is immersed in water. This relationship is complemented in humans by social elements of a higher level (reason and relational correction). This is the part of the stick that is in the air.

What is described above are the relations that can be both positive and negative. Let us make a step now to love as probably the strongest positive relation of which humans are capable. Something like love happens when we are hit, at the first meeting, by "a bolt from the blue". We become immediately aware that this person meets most of our imaginations (requirements for the attributes) that are expected from our dream-world partner, the partner we cherish in our heart as the one we are looking for. The process of an instinctive as well hormonal assessment of the person goes on at this instant as well as the process of the social check (level of the reason and relational correction) that enhances the instinctive and hormonal assessment. On the rational level, men do not consider each woman primarily as sexual objects. This is directed by morals and learned social customs. However, in the background (subconsciousness), he **always** views them as such and it is now important whether the sexual aspect prevails and the other aspects (social ones) are secondary.

For example, at a job interview (a question of resource provision), the first meeting with a woman boss carries a highly social context (not a sexual one). However, at the subconscious level, there is always a concurrent sexual aspect and these two aspects can be quickly reversed. The same situation may be with women and their man bosses. This reversal is tightly linked to the relational table — on the job we are driven by the instinct to provide resources but the sex instinct can quickly get the upper hand. From the very instant when the particular person becomes an object of sexual interest in our head (the parameter "is this human of sexual interest" in the relational table is set to "very much so"), we cannot dispose of this aspect. We can blame the power of the sex instinct that takes control over the relation. It is obvious that a man sees an adversary in the man boss, and in a similar way a woman considers her woman boss an adversary.

Infatuation happens when we rate an individual as the perfect one by our relational requirements (parameter values in our relational table); we have a strong desire to live with him or her. Since "no one is perfect", this may exceed a certain threshold in checking out this individual, and we may pardon some small faults as "he or she is otherwise perfect". ☺ It is even highly probable that when we exceed the perfection level of this individual we may stepwise start suppressing the negative parameters ("rose-colored glasses").

10.3. When infatuation moves to love

Infatuation as the state of perception of a partner as the "ideal mating partner" transit over to the love in the moment, when the simple (instinctive) approach creates rules in the brain, which "map" the existence of the (ideal) partner into an image affecting our acting and thinking in such a way that the partner and the image commute into one individual, of course in the model of the loving person (not only the infatuated one) only. Extraordinary strong love shows, when the beloved partner is generaly privileged in acting and thinking of loving person. It is quite sad when beloved partner does not return love and starts to misuse love for base purposes. I am sure that you know such behavior in your circle of acquaintances.

Sometimes love can develop slowly because we, for some reason, delay its development (self-preservation instinct, reasonable doubts, et cetera) and we give it a free run later.

CHAPTER 11

Faith, God, atheism, and the like

11.1. Having faith and be a believer, there is a difference

In this chapter, I will try to present a simple solution for one of the fundamental issues of humanity. The issue is believers versus non-believers, and the explanation follows by using computer science. This approach finds that both the believers and non-believers share the same principles of word view and this also explains why of the best human brains of present and past time turn to God in later stages of life. These humans cannot be under suspicion that they believe in God on high heaven who directs everything. This fact stole all my sleep until it dawned on me: these most clever people in the world understood their limits, and this is the consequence.

When we take a look how religions perceive God, we can see that they usually perceive of Him as a Being, though imaginary only. My efforts are to unite the understanding of God in different religions and cultures. This Being is perceived as a being without a material body. Otherwise, It cannot possess some of the characteristics that are assigned to God. I will try my definition of God:

God is whoever knows what neither I nor any other human in the world (on earth), know.

This definition covers almost any imperfection, any non-competence (as an un-acquired competence) and the inability of humans to cope and understand. It is not the case of only a piece of information to the ordinary human that somewhere there beyond our reach and understanding is Someone (Something) who (that) knows all and is so capable. It is also about us to be able to describe and denominate an area that obviously exists and works but exceeds our capabilities. This area concerns the elimination of the unknown and the emptiness. The existence of the unknown horrifies us since we cannot work with it and do not know what lies ahead and from where danger comes. Creation of an understandable model of "what I am not able to grasp myself" diminishes human stress and gives us peace.

However, the creation also gives space to more clever and more capable people to control other individuals through the detailed specification of the unknown. It is quite essential to make the ordinary human believe that Someone knows the unknown better and better understands what is going on and what is wanted. An apparatus of mediators should usually be formed for this to happen. This mechanism should provide for the communicators (prophets, son of God, Pope, and others) between God and the ordinary people and these mediators must be sufficiently beyond our grasp for the psychological mechanism to function (miracles, for example).

Faith in a God, who is a part of nature (as with the Indians, Polynesian tribes, and others), is an independent category. This mindset is, in fact, more direct and more rational than the attitudes mentioned above to personify God – an explanation follows below.

We introduce now several notions (definitions) that will, I hope, express with sufficient precision the material notions linked to the existence of God. They may have different names but will facilitate our understanding. In some cases, the notions just reiterate what some religions already contain. Let us go for it:

God = nature, the whole world around us (including the one we do not even know about ☺). When the leading physicists (Albert Einstein, Niels Bohr, Wolfgang Pauli, Stephen Hawking) spoke about something related to God, they always spoke about nature, which, however, is endowed with will[61]. In this way, they in fact designated the laws of nature as God.

Faith = finding/understanding of own computational and informational limits in the way that we concede that there are facts not known to us, and we will most likely never know them, but we can work with them as assumptions. The realities held in the faith (faith in something) are taken axiomatically (an axiom is a premise that has no proof, and any discussion about it does not make sense). However, a constant critical look at the axioms' specification is acceptable minimally until we find enough information and facts for a discussion (about so far established facts) and consideration about the validity of an axiom.

Religion = (1) an experiment to give people a "social system" (operating system of people), in other words, a "network (internodal) operating system"; (2) an instrument to control other, less capable, people. Both roles may be implemented concurrently, and item (1) is practically always positive for society while item (2) can be both positive (leadership by a just and fair leader) and negative (unwished leadership of type "slavery").

---

61 The most famous statements:
God does not play dice. (Albert Einstein)
Not only does God definitely play dice, but He sometimes confuses us by throwing them where they can't be seen. (Stephen Hawking)
Don't tell God what to do with his dice. (Niels Bohr)
I cannot believe God is a weak left hander. (Wolfgang Pauli)

11.2. Everybody has some faith even if claiming otherwise

Let us attempt again a computer science architect's special view of humans as systems, responding to the surrounding as an environment where humans function and to which they react. Imagine a system containing all the components and information and also the procedures of handling these components and information.

It is of course very difficult as we are, in fact, designing a new description for the world ☺ ; it is, however, necessary to understand what "faith" means to humans. We can simplify these ideas by imagining the system as a flat surface – large and unbounded. We can now imagine the knowledge and experience of a given human, a part of this set, as a circle on this surface (when we put it on paper, then this circle is a part of the whole surface, which we "cut out" for our own usage). This circle then represents the knowledge and experience of the given human and the rest outside the circle stands for what the human does not know nor can he do anything about. ☺ The brain of this human can thus process the information and methods inside the circle (bubble) as if he or she acquired them or verified them and took for own ones. Everything else around, if the brain has to work with it (and every brain of every human must, in fact, do it), is something to be taken as fact (if this human does not prefer to ignore it), even though we do not know if it exists at all, mostly we cannot touch it, we only have a mathematical interpretation of the fact, or procedure in the sense that it exists and functions.

This information and procedures stand **outside the circle of knowledge and experience of the given individual** and constitute something in which he or she must believe (it is the prerequisite for further work of brain) – it is the **belief** of the individual. These "surroundings" create, in fact, the environment for the internal rules (see chapter 4.2.) in the human brain. In this case, that is the "global environment", i.e. the environment outside the brain. We need this global environment to find a solution of a task under the conditions of the outside world of the individual.

Thus, it follows from what we have said that every human being must have faith — believe in something, take it as a prerequisite (axiom). This is not only normal, but it is also a very ingenious method (a mathematician would crack a smile at this simplification) to account for what we do not know, or about what we are not sure. It is a kind of a basic prerequisite to working with what we do not know or cannot do.

Since the capabilities of different humans vary, we can describe these differences by the size of the circle (bubble) — whoever has four times as much knowledge or experience, will have the circle with twice the diameter (we are on a plane — twice the diameter means four times surface area). Who has nine times greater knowledge and experience, will have the circle of thrice the diameter.

The capabilities of the most stupid human to the greatest genius in the world can differ by multiple orders of magnitude, that is 10x, 100x, perhaps even 1,000x. However clever the most clever human in the world can be, the size of his or her circle can never be the whole surface representing the Universe. Of course (figuratively), the future most clever individual in the world would increase the area of the circle to be the largest one. He or she defeats one ignorance after another and turns them into knowledge. However, such a life-long "circle enlargement" must sometime finish. We grow old, tired, senility steps in, Alzheimer's disease advances, and so on, and the circle enlargement stops at some time. This clever individual then realizes at that moment (he may have known it for a long time already) that somewhere outside the circle there is something "superhuman" which exists, but he or she has no chance to learn it. At this instant this super-clever individual, a thorough proponent of materialism, may come to be aware of something we may denote as God. In reality, it is the nature (the Universe), and God is just another name for the same.

Given the above, such an attitude is quite normal, self-reflexive and humble. What matters is that this super-clever human will not turn to religion — faith in God (or nature) has nothing in common with a religion.

A FEW CRAZY IDEAS

Religion is a method to control people using God (see the definition above). It is essential, in the case of religion, that people (who are to be controlled in the future) may believe, that someone has better information and thus the right to say what others should do.

CHAPTER 12

How the national wealth originates

12.1. How does Ronald Reagan relate here?

Ronald Reagan, President of the United States from 1981 to 1989, expressed this interesting idea, in an answer to a question on humanism and solidarity:

"I'm sure everyone feels sorry for the individual who has fallen by the wayside or who can't keep up in our competitive society, but my own compassion goes beyond that to those millions of unsung men and women, who get up every morning, send the kids to school, go to work, try to keep up the payments on their house, pay exorbitant taxes to make possible compassion for the less fortunate, and as a result have to sacrifice many of their own desires and dreams and hopes. Government owes them something better than always finding a new way to make them share the fruit of their toils with others".

I found this thought after I had realized a simple truth I am introducing two paragraphs below and which, in my view, represents the very foundation of success and wealth of each nation on the earth. I have borrowed this thought of Ronald Reagan to explain better why I think so. This thought of Ronald Reagan is, I dare say, the thought of a genius.

12.2. What is at the beginning?

If we start to inquire about where there lie the very first foundations of the fact that some nations prosper, and other barely survive (we do not speak about communities where one or a couple of individuals appropriate most of the nation's results), one of the possibilities is the idea which occured to me:

A wealth of a nation depends in principle on whether the majority of the people are willing to get up in the morning and do something meaningful.

By "getting up in the morning and doing something meaningful" I mean going to work whatever this work may mean – hard labor on a building site in all weathers, or operating a supermarket checkout counter and clicking away for a day at the keyboard, or, perhaps, sitting for a whole day in a warm office as a rentier thinking where to invest.

It is quite essential whether something forces the given individual to make this above step. It may be, for example, to provide something for heating and thus ensuring the survival of the family or him or her (so that they do not freeze to death).

In a similar way, another essential need appears – food. The individual must muster the strength and find something to eat. The amount of provided food will greatly affect whether the individual just makes it or lives comfortably (I think it in the sense that he can afford other offsprings since they will not die of hunger).

Let us take a look at those regions where nothing forces humans to go and provide heating but food "only". It could be, for example, Africa.

There is enough warmth which is fine, and they do not have to do anything about it (neither at present nor a million years ago). If, moreover, such a human in Africa lives close to a sea or lake rich in fish, he or she need not exert too much effort to provide food for the whole family and himself or herself. One hour of effort in a day, in fact, is quite sufficient for the survival of himself or herself and the family. This is valid, of course, for the period of stability (primarily of the weather) and under the condition, that the population does not start to grow out of proportion so that the activity (better: non-activity) of the population for the provision of food is out of line to the number of inhabitants, i.e. the famine does not arrive.

Quite a different situation exists in Europe, both at present and in the time of the Ice Period (tens of thousand years ago – these periods appeared on a regular basis). A human and the family had to fend for themselves quite hard to survive the winters and have enough to eat. There is not enough sources in winter to heat, nor enough animals that are often hidden in their dens to survive. The energies for the survival of such humans are in the multiples (by orders of magnitude – easily 10x to 15x) of those who live in a warm region for whole years.

This is a well-known result in the evolution of humans – although Homo Sapiens evolved in Africa, his further substantial evolution (mental, social, then technological, cultural, et cetera) happened "more in the north". The same also holds for the colder regions south of the equator, but humans in Africa had only one direction where to go without having to cross the sea for thousand of kilometers. This direction was via the Near East to the north (perhaps even through the Strait of Gibraltar. However, at those times humans were unlikely to swim across this strait of tens of kilometers except perhaps by accident, when a few individuals in a storm drifted on a tree log toward Europe).

Conditions in the north, however, ceased to be so favorable as in Africa, and so people had to make much more effort to survive (I intentionally leave unanswered the question why they went originally northward when the conditions were not favorable). Thus, the further evolution and development of humans occurred through greater efforts necessary for survival. This is fully understandable, and there is nothing earthshaking about it; it is, however, a foundation for our consideration mentioned at the beginning of this chapter. It is quite interesting that most of the nations that got rich by making an effort are in the moderate zones of type "Europe" and this may be caused by the life conditions force people to work regularly to survive. This is in contrast to other nations that only sell raw materials (which they did not create).

The development in our civilization has done away with the daily chores of providing food and heating. However, wherever the particular nation (this characteristic is usually bound to the nation as a whole as it is tied to education and morals) has not relented in these efforts, it stands much higher in the table of competitiveness and the overall wealth. This is, again, in contrast with those nations where the attitude to life content is not so intensive. I would like to say that even the small fact that circumstances (de facto, the social norms of the nation under scrutiny) do not require most of the citizens to engage in a sensible daily activity, does affect in a negative way the wealth of each individual (of course, on average). I hasten to remind that this is true with a democracy though in the initial stage, and not a dictatorship (even hidden ones). In this dictatorship, a few individuals lead a fabulous life and the rest of people barely subsist. In such a case, the wretched individual members of the nation have no motivation to work and strive only to survive. There are exceptionally so-called enlightened dictatorships could be in a way transitory phases toward democracy.

Imagine a case that may seem to people from Europe or the U.S.A. rather crazy but it can be a lovely example of "taking off the ground". A populous country with an undeveloped industry (there are many such examples) has not enough electricity to further develop the society. It does not even have sufficient investments into any generating plant – there are some budgets but not enough for a classical investment into a modern generating plant to be imported from an industrially developed country. What to do? Lament and beg for "aid"? No. Those who would like to do something serious, here is an example:

A million people (a fraction of the total population) equipped with some primitive tools (at least a scoop of some shape, if there is no money for a shovel and wheelbarrow) can go on carrying clay in hands for a whole day according to a simple project (for such a project, money is available or can be borrowed), all under the supervision of one or just a few experts (for them money is available or can be borrowed) and we will have an earthen dam, we can put in some cheap electricity generator there (sometimes the developed countries get rid of old technologies and give it away just as a free takeaway, as the advanced economies can thus innovate and be more efficient). And, lo and behold, suddenly we have a source of electricity here, at least a minimal one, but this is MULTIPLE times better than no electricity at all.

The only thing needed for something like this to happen is: THEY MUST FEEL A WANT. A want to go and work daily at a dull work and do it sufficiently long (one or two years). A want to do it for board only - the same they did before the "project". Before the project they hardly subsisted, they subsisted during the time of construction, and they have moved higher – they have electricity. It may, however, be that nothing forces these people to do anything at all – they have sufficient food (fish from the sea, fruits and vegetables grow wild without any husbandry – bananas, mangoes, et cetera) and so they have no basic need for any activity to survive. Our phantasies that they want to work so that they may "advance" could be unfounded in this case. The development of such a social system, in this case, slows down and hence stagnates.

There is nothing complicated to implement such a scheme. It can be however implemented under one condition only – there must be lots of people who get up early in the morning, even when they do not feel like it, and start to do something making sense.

Please take this example only as a model example; but why could it not be real? It is absolutely not necessary that it would happen in a highly developed country. It would be rather quite surprising.

12.3. Too naive? Too unrealistic?

Fairly recently we have seen banks in some poor regions to be successful with the so-called microloans. What is a microloan? 500 USD? 5,000 USD? No! It is, for example, 5, 10 or 50 USD. What kind of nonsense is this? Is this loan for a lunch or what? No way — they will buy a shovel, for example, a wheelbarrow, or some other primitive tool — an item of means of production! It may sound incredible, but those people do not have money even for this "investment", and so they must borrow. Surprisingly, this way of doing business causes a boom in some less developed regions.

This attitude may remind those of us who like mathematics, in particular, of the integral calculus (concerned with adding up a huge number of infinitesimal elements — an integral by dX over a given range). This similarity is NOT just a pure contingency. ☺

Another example is the initiative of Mrs. Bernice Dapaah from Ghana (Africa)[62]: she makes bikes from bamboo. It is obvious what such an approach brings: plenty of cheap local material, the material is ecological, extraordinary endurance of the material (bamboo is the material of future), et cetera. The project is economically successful and the bikes are exported to the neighbouring countries. The producer donates the bikes as a charity activity. Mrs. Bernice Dapaah from Ghana is a genius.

---

62 See https://www.facebook.com/ajplusenglish/videos/672599906214853/?theater

CHAPTER 13

Epilogue

13.1. What to add?

If you have read up to this place, it is a great honor for me, as apparently my temerarious thoughts have not quite put you off.

If, from time to time, some people would argue among themselves about my points (in a constructive and friendly way, of course), I will be very happy as I have inspired them for deeper thinking.

It could happen that my ideas will only survive in such a form: anyone who would get some knowledge of them would amuse his or her immediate circle of friends by speaking about the biggest idiocies he or she has heard about. However, perhaps some individual so disposed as I, might say: "Hey, wait a minute – there could be something there" and then he or she would spend some time on them and in the end would refute the ideas - then I would be even happier. This is how searching for truth should look like.

13.2. Immodest wish

If, however, it would turn out that at least one of my ideas is proven true, my dream will have come true.

And that has been my wish in writing this book.

CHAPTER 14

Glossary

Term	Explanation of the term (source / sources)
abstraction	**abstraction** is the ability to work mental (in-mind) with nonmaterial objects – ideas, areas, categories and features – as the opposite to sensory/ material processing when I can touch or smell the material object I am processing. The ability to abstract is occasionally limited just to the hierarchy of objects and to the relation among the abstract objects. The common basis of abstraction processing is the fact that everything is a thought model, i.e. a nonmaterial model is created, which is subsequently processed using information and mathematical tools. Abstract objects can relate to a material object or can relate to other abstract (nonmaterial) object. It is apparent that humans started with abstract modelling of material objects (he or she imagined it and made 3-D operations on it – for example, they imagined where the quarry moves and what they as the hunters will do) and later he or she started to process purely nonmaterial objects, which never existed before (philosophy, science, et cetera.). (**source**: an own definition of the author based on the general understanding of this term in philosophy, mathematics and computer science)

awareness	awareness (vigilance) is -
	(1) the ability to perceive, feel, or be conscious of events, objects, thoughts, emotions, or sensory patterns. At this level of consciousness, sensory data can be confirmed by an observer without necessarily implying understanding. More broadly, it is the state or quality of being aware of something. In biological psychology, awareness is defined as a human's or an animal's perception and cognitive reaction to a condition or event. (**source**: www.wikipeda.org – English) (2) the state of readiness of the organism for reaction to events. It is a basic operational state of consciousness, when all important subsystems are ready for activity and for willful (rational) reaction to events. The opposite of awareness is subconsciousness, sleeping, unconsciousness or coma, of which every that state has its own specific role. Awareness lasts roughly 16 - 18 hours a day. The human acquires during awareness the information, which is subsequently processed by subconsciousness during sleeping. The medium-term memory is released after that processing and the brain is ready for the next period of awareness, i.e. for next conscious activity. (**source**: www.wikipedia.cz – Czech, slightly edited by the author)
IT	computer sciences, information technology, informatics

intuition (anticipation)	intuition is - (1) a phenomenon of the mind, describes the ability to acquire knowledge without inference or the use of reason. The word "intuition" comes from the Latin verb *intueri* translated as *consider* or from late middle English word *intuit*, "to contemplate". Intuition is often interpreted with varied meaning from intuition being glimpses of greater knowledge to only a function of mind; however, processes by which and why they happen typically remain mostly unknown to the thinker, as opposed to the view of rational thinking. Intuition has been the subject of discussion from ancient philosophy to modern psychology, also a topic of interest in various religions and esoteric domains, as well as a common subject of writings and is often misunderstood and misinterpreted as instinct, truth, belief, meaning and other subjects. The right brain is popularly associated with intuitive processes such as aesthetic or generally creative abilities. Some scientists have contended that intuition is associated with innovation in scientific discovery. (**source**: Wikipedia)

	(2) the answer of human's subconsciousness to the question of consciousness. Consciousness does not know where the answer appeared from, i.e. it cannot specify the source of information, so it perceives that the answer appeared all of a sudden by itself. Intuition contains usage of the whole memory and experience of human individual as the subconsciousness has access to that informatin. Even more, intuition contains usage of the experience of ancestors of this human individual as it is transferred in some way from parents. The result is usually vague (it is not YES or NO), so the consciousness understands it as a "feeling". The answer could be delivered in the form of operational environment (parameters) for a conscious activity. (**source**: the author)
telepathy	**telepathy** (from "distant" and "feeling, perception, passion, affliction, experience" in Ancient Greek) is – (1) is the purported transmission of information from one person to another without using any of our known sensory channels or physical interaction. The term was coined in 1882 by the classical scholar Frederic W. H. Myers, a founder of the Society for Psychical Research, and has remained more popular than the earlier expression thought-transference. There is no scientific evidence that telepathy is a real phenomenon. Many studies seeking to detect, understand, and utilize telepathy have been carried out, but no replicable results from well-controlled experiments exist. Telepathy is a common theme in modern fiction and science fiction. (**source**: Wikipedia)

	(2) the unidirectional or bidirectional transfer of information between two (or more) human brains (it apparently exists at animals, too). It can be a simple item of information (basic signal – a remote "nudge") or a complex information (idea, name, relation, health status of the body, et cetera). Such a transfer could be done using a sense which has not yet been discovered. This probably sense could use a sensor which is very small (tens or hundreds of molecules) with an extreme sensitivity similar to the sensitivity of the retina cell with energy sensitivity of the one photon, i.e. 10^{-19} Joule). The author presents such a theory in this book that such a sense exists and proposes a possible technical solution of it. This understanding of telepathy is strictly separated from any supernatural phenomenon (see the basic assumption in chapter 1.); for example future forecasting / prophecy (foresight, et cetera) which is one of the unclear tasks in philosophy and / or computer science view because of necessity explication of causality (chain of causation). (**source**: the author)

consciousness	consciousness (to know) is the fundamental attribute of an aware human being and in limited degrees also of other animals. Consciousness integrates various mental activities (perception, feeling, speech, memory, attention, et cetera) and controls them to a certain extent. It is the source of will and volitional actions, decision-making, and hence also the focal point or center of the human person. We lose our consciousness in sleep, in serious diseases and brain trauma, it is thus dependent on the brain. In our modern thinking, it replaces the obsolete notion of soul or "self". Consciousness in medicine is the opposite of unconsciousness, sleep or hypnosis. (**source**: Wikipedia).
entropy	entropy is an authority on the "degree of disorder" or "degree of indeterminacy" of a given system. (**source**: Czech version of Wikipedia).

CHAPTER 15

Peer reviews

Viktor Zeisel, a specialist on life, folk philosopher, part-time ICT man:

I have known Petr a damned long time – I'd say from 1991. I know his professional career and it arouses respect. That is why I cannot be unbiased. I do not agree with all his opinions and theories, but he always forces me to think about what he does and says. Then I can confront him with my views and opinions and argue against his ideas.

I thought quite often about the same topics as Petr, as I am a man with technical education and approach, too. I regard mathematics as the king of sciences, too. Mathematics looks and measures at the whole world functionality, especially that of communication.

Petr had the patience and enthusiasm enough to collect the information, to find relations between them (though theoretical only) and to compare individual theories. Furthermore, he deliberately underwent the risk of being ridiculed. It can appear from the people who know much more on the individual topics and they will consider Petr's assumptions and speculations as naive, unfinished and, maybe, misleading. Furthemore, it can appear from the people who have limited imagination ability, knowledge and passion to find relations where they are not obvious - the relations which disturb the ease of routine clichés and which force you to doubt about agreed truths.

There are definitely many people, who are much more competent to deal with these topics, especially with each of the individual discussed topics. I appreciate the fact, that I was able to understand the individual thoughts, as they are enunciated explicitly enough to enable to grasp them, criticize them, modify and develop them by anybody who is strong in fantasy and imagination. The thoughts are not presented in dogmatic way, he does not consider to half-truths as proves and he does not prevents us to use our own approach and understanding.

The reading was funny and instructive. I acknowledge I appreciate the foray into computer theory as I nearly forgot several things in the flood of social networks, clouds and graphical gadgets.

Was there in the book something which held my interest? What absolutely grabbed me? Something I will return to? Yes — it is the theory that the process "to get an all-new idea" could be realized by the solution of a system of differential equations.

I asked myself many times what to do with creativity, creative inventions, non-conventional approach? It is necessary to put up the relevant differential equations, check their number, check if some equations are not in fact identical. And, of course, to define an effective method of solution. Fortunately, the human brain, consisting of neurons connected by synapses (you see — the book left a marked impact on me), is much more powerful than any computer and so our common sense is to be enough.

Many thanks for the opportunity to be among the first readers of this book, in its draft state, before the book reviews will remove some absolutely controversial passages and opinions.

Marek Šetina, heart surgeon:

A characteristic of the work in one sentence could be as follows: clever book for clever people anxious to learn, who enjoy a novel view of the world.

A FEW CRAZY IDEAS

The book is brimming with ideas. The range of knowledge of the author is admirable and it covers many fields – mathematics, cybernetics, history, astronomy, medical science, genetics, anatomy, psychology, natural sciences, …

The fact that the author uses information from different branches of science has both the advantages and disadvantages. On the one hand, it necessarily leads to shallow or too schematic description of a problem. On the other hand, it enables the author to remain at the "big picture" level and not to be overloaded by too many details. The combination of knowledge from quite different areas gives the author more creative freedom and it enables him to present an unconventional view – in author's words: a "crazy view". The author does not bother too much about giving scientific proofs.

Mankind needs visionaries for its progress who hover with their ideas somewhere in the realm of fantasy and have no patience with a tiresome search for proofs. However, mankind also needs courageous conquerors and pioneers, who believe in their ideas and venture out to explore the unknown. These are the scientists who eventually deliver the proofs after in-depth research if the theory or vision are correct. Even scientists are wrong, sometimes, so it is not rare that today's truth is tomorrow's falsehood, and vice-versa. Last but not least, we need the implementors to bring an idea into reality, of course.

The autor is definitely a visionary and, maybe, he provokes, too. His goal in not only enlightenment or presenting some statements or ideas, but it is an attempt to incite readers to discuss. He is aware that it is the conflict of ideas what moves mankind forward, the necessity to defend a controversial opinion, polite disagreement and unconventional and different view. Such an approach enriches mankind.

I believe that the book fully meets its intention. It is readable, written in fresh pace and light spirit. It brings for many facts and inspiring ideas.

Martin Kobr, almost my double, but he still believes he is more clever:

Me, the peer reviewer, declare that the book (state as of October 2015) fascinated me, it brought me a lot of fun, it delighted me many times and it made me sad a few times.

I hope that at least some of my humble comments will improve the quality level of the work, which is high enough already. ;-)

Well done, author! I wish all the best to the book! Good work, engineer. Howgh.

Petr Rektorys (born 1959, in the Czech Republic) received his education in civil engineering at the Czech Technical University in Prague. However, he pursued his professional career in engineering for only three years after graduation as his heart tilted elsewhere – to computers and information technology. Computers and computer science have been his hobby since the age of seventeen when the first programmable calculators from HP and TI appeared on the market. This hobby, together with his inclination towards mathematics as the fundamental tool of information technology, took a full hold of him. In his ensuing "course of life", he was engaged with the application of computers and modern electronics as encountered in the daily life of normal mortals (not professionals). He was reading lectures on these subject at conferences already during his studies as well as publishing in technical papers. He subsequently extended his professional career into the implementation of computers in Enterprise Resource Planning (ERP). For his pure personal pleasure he has been exploring for almost his whole life a spate of philosophical questions (life – its purpose, origin, development; the world and the universe and the laws of their workings, etc.), as well as the possible technological solutions involved (in applied mathematics, physics, astronomy, biology, genetics, and mainly computer science as the underlying tool for these fields). As early as circa 1980 he elaborated the topic of understanding proverbs in terms of differential equations which form the basis of human mental activity (brain function). After 2010 he developed the concept of using the recordable part of the genome (not yet confirmed) to explain both the development of species and the previous lives of individuals. This book is his first book and was planned to appear only after sufficient material had been accumulated. The main effort of the author in this book is to merge the views of experts in different fields. As a rule, these experts do not understand each other because their fields are so much apart. Such an inter-disciplinary approach is generally considered to be the key for further development of scientific research towards new fascinating discoveries. This book originated in order to stimulate discussions on this approach.

The reviewers's opinion:

Me, the peer reviewer, declare that the book (state as October 2015) fascinated me, it brought me a lot of fun, it delighted me many times and it made me sad few times. Well done, author! I wish all the best to the book! Good work, engineer.
(Martin Kobr, nearly myself, but he still believes he has better ideas)

The characteristic of the work in one sentence could be as follows: clever book for clever people anxios to learn, which enjoy a novel view of the world. The book is brimming with ideas. The range of knowledges of the author is admirable and it covers to many fields — mathematics, cybernetics, history, astronomy, medical science, genetics, anatomy, psychology, natural sciences, et cetera.
(Marek Šetina, heart surgeon)

Was there in the book something what still holds my interest? What absolutely grabbed me? Something I will return to it? Yes - it is the theory that the process "to get an invention" could be realized by the solution of the system of differential equations.
(Viktor Zeisel, a specialist on life, folk philosopher, part-time ICT man)

Discussion panel:
www.aFewCrazyIdeas.com

The contact:
aFewCrazyIdeas@gmail.com

ISBN-13: 9781537759685
ISBN-10: 153775968X

www.ingramcontent.com/pod-product-compliance
Lightning Source LLC
Chambersburg PA
CBHW021426170526
45164CB00001B/110